War *of the* Gods

War of the Gods

Alien Skulls, Underground Cities, and Fire from the Sky

Erich von Däniken

This edition first published in 2020 by New Page Books, an imprint of

Red Wheel/Weiser, LLC
With offices at:
65 Parker Street, Suite 7
Newburyport, MA 01950
www.careerpress.com
www.redwheelweiser.com

Copyright © 2020 by Erich von Däniken

All rights reserved. No part of this publication may be reproduced or transmitted in any form or by any means, electronic or mechanical, including photocopying, recording, or by any information storage and retrieval system, without permission in writing from Red Wheel/Weiser, LLC. Reviewers may quote brief passages.

ISBN: 978-1-63265-171-6

Library of Congress Cataloging-in-Publication Data available upon request.

Cover design by Kathryn Sky-Peck
Interior by Maureen Forys, Happenstance Type-O-Rama
Typeset in New Baskerville, Trade Gothic and Caslon 224

Printed in the United States of America
LB

10 9 8 7 6 5 4 3 2 1

CONTENTS

Letter to My Readers . . . . *vii*

CHAPTER 1: Alien Skulls 1

CHAPTER 2: Veil Dance around the Pyramids 37

CHAPTER 3: The Exploded Planet 85

CHAPTER 4: Skeletons—Not from This World. 165

EPILOGUE: And Yet It Moves!. 197

Bibliography *199*
Image References *211*
About the Author *213*

LETTER TO MY READERS

DEAR READER,

In this book, I present new findings! But it is only possible by building on previous experiences. Every reader of my books knows what a great dolmen is—but no one has ever heard of the dragon houses in Greece. In fact, great dolmens and dragon houses came about for the same reason, and we have all read about the underground cities in Turkey, but no one has drawn parallels to the so-called primary caves and Erdställe (earth stables) in Austria and Germany. Why did our Stone Age ancestors dig burrows for themselves underground? And why did they do that worldwide and millions of times? What was the force that drove them to do this?

Between Mars and Jupiter lies the asteroid belt, also called the planetoid belt. How did it actually form? Could the hundreds of thousands of smaller and larger chunks of rock be the remains of an exploded planet? But heavenly bodies do not just explode like that. Are there references to a star war between gods in ancient literature? Was that the reason people sought protection from the cosmic projectiles? And where did the survivors go?

For generations, "deformed heads" have become world famous. These are "elongated skulls" pointing backward. They exist in the tens of thousands, and up until now we have been able to clearly prove how they came about. But more recently, skulls have appeared that no longer match the expected type. The same is true for some skeletons. What is going on here?

This book tries to explore these findings. Parallels must be drawn to previous books. The new findings are the culminating point of research reflected in my forty-one books so far.

With warm regards!
Erich von Däniken

War *of the* Gods

CHAPTER 1

Alien Skulls

FEBRUARY 23, 1988. It was to become an unlucky day. I had led a tourist party into the South American highlands. We arrived from La Paz, Bolivia, landing in Santa Cruz de la Sierra. At around 4,000 meters above sea level, La Paz is the highest civil airport in the world from which jets are still permitted to take off. Santa Cruz de la Sierra, our destination, however, was located at an altitude of 3,500 meters lower than La Paz. In the highlands, some had complained of breathing difficulties. Here, at 437 meters above sea level, the group was thrilled. We had all the oxygen we needed. At around five in the afternoon, we met for a drink at the swimming pool of the Holiday Inn.

"Erich, did you see Rob?" his wife Julia asked. "He's not in our room, and neither is his bathing suit. He has to be down here somewhere." Julia and Rob came from

Holland. This trip to faraway South America was their honeymoon. Rob was an imposing figure, six-foot-five-inches tall. Athletic, slim, fit. He had often told me about his extensive cycling tours. I got up and strolled around the pool. I looked for Rob. Had he perhaps left the hotel in his swimming trunks to buy something at the small stands in front of the main hotel entrance? Was he in a restroom? Our group included three young men. I asked them to search the whole facility for Rob and to check every restroom. To no avail. Rob remained missing.

Suddenly a woman screamed down from the second floor balcony and pointed excitedly to the swimming pool. It didn't look clean. No blue, transparent water. There, at 2 meters below the greenish, gleaming water surface, lay a human body with outstretched arms. We maneuvered Rob to the surface, heaving him out of the water. But the CPR performed on him came too late. A doctor diagnosed a heart attack. Ironically, Rob, the athletic type, had not coped well with the difference in altitude. We all took care of Julia, but she was surprisingly accepting and brave.

We ate dinner from the buffet. The group remained very quiet; everyone was talking about Rob's death. Julia wished that her husband's body would be transferred to the Netherlands. I promised to arrange this, although at the moment I had no idea how to make that happen and, specifically, didn't know what that would cost.

Then, keeping a friendly distance, the hotel director stood before me. I knew him from an earlier trip. He was a well-educated man who had also studied for four years in the United States. Mr. Antonio, as everyone called him, asked me to come to the front desk. There was a black man in a dark suit, wearing white gloves. The suit did not fit the face. I thought he was from a funeral home and wanted to discuss details with me concerning the transport of Rob's body. But Antonio told me something else. The man with the white gloves was the servant of a very rich man from Colombia. The stranger would like to meet me in person. His servant would take me to him and later back to the hotel.

"But Antonio," I replied, "my group is sitting in the restaurant. I cannot leave them alone." Antonio pleaded with me to get into the black man's car. I would not regret it, he insisted, and he assured me the rich man was also a major shareholder of the Holiday Inn. In addition, the hotel would comp all of tonight's drinks for our entire group. That was an excellent offer. "What's the name of the stranger, and what does he do for a living?"

Antonio squirmed. He said he wasn't at liberty to answer these questions. I would learn everything at the house of the *hombre rico* (rich man). I went back to the group, told them about my plans, and said I would probably be back at the bar in about an hour.

A dark Mercedes limousine was parked outside. The black man with the white gloves pointed to the back bench. Affixed to the back of the seat in front of me was a little rack to hold things. It held three bottles, each one containing a different type of exquisite whiskey; some mineral water; ice; and two glasses. I did not touch anything. Slowly, the vehicle drove up a hill. At the bottom right, I saw the lights of Santa Cruz de la Sierra. Then the chauffeur turned onto a bumpy, unpaved road. We drove through a grove, and I began to wonder if the whole thing was really a kidnapping. I did not carry a weapon on me and would have been helpless against my kidnappers. Then I abandoned my crazy thoughts. It was clear that my group of tourists in the hotel knew that I had gotten into the Mercedes on the director's recommendation. And anyway, I did not see Mr. Director as a crook, and moreover, if it were really an abduction—why was it carried out in such a complicated and luxurious way?

Finally we stopped in front of a gate. After we had passed through it, two native South Americans closed it behind us. As I got out, I realized—all men were armed. In front of us was a one-story building with lots of green plants on the outer walls. Inside I encountered a spacious hall and several beautiful ladies clothed very sparingly, who smiled at me seductively. Did I end up in a bordello? Finally the "rich man" walked up to me.

He seemed like a nice guy, with dark, frizzy hair, lively eyes, and a well-groomed mustache. Impeccable teeth as well. The lower lip was slightly wider than the upper lip, and his chin had a dimple in the middle. Somewhere I had read that a dimple means generosity. The stranger wore jeans and a silk, beige shirt, and his left wrist was adorned with a diamond-studded Omega watch.

The stranger welcomed me with "Señor Erich" and a strong handshake. He said his friend Antonio from the Holiday Inn had told him that I was the guide of the tourist group, so he wanted to take the opportunity to meet me personally. Without exaggeration, he apologized for the unusual "kidnapping" and assured me that one word from me would be enough to get me back to the hotel immediately. Could he be of any help to me?

I informed him of the unexpected death of a travel companion and said I did not know who was responsible for transporting a corpse from Bolivia to Amsterdam.

"Do you carry an American Express credit card?" he asked. I always carry one in a small inside pocket of my jacket. I gave him the number.

"Don't worry about the body," the host said. "For the flight to Europe, the blood must be extracted and replaced with formaldehyde. That happens here at the hospital. The transport is organized by American Express." That's exactly what happened. Three days later, Rob's body landed in Amsterdam.

He had read some of my books, the lively stranger assured me, and he wanted to show me something unique.

First, I was served cold champagne, and sandwiches were passed around. The stranger said I should call him Pablo and he could support my research, especially in Bolivia and Colombia, where he owned quite a lot of land. I inquired about his job, and he said he was a car dealer and also a rancher. We talked about the visit of aliens on good old planet Earth, and Pablo complimented me and assured me that I was one of the few contemporaries who understood the big picture. After half an hour, I told him that my group would be waiting at the Holiday Inn. Pablo asked for a few minutes' patience. Then two native South Americans came to our table. They carried something that looked like a very long shoe box with hinges and set it on our table. Pablo flipped down three sides of the container, pointing inside. "What do you think, Señor Erich?"

"A deformed skull," I said, baffled. "You can find them in various museums, including the Anthropological Museum of La Paz."

"I know," Pablo nodded. "But this is the head of an alien."

"I beg your pardon? Why do you think that?"

Meanwhile, servants had set up a portable screen and put a Kodak slide projector on the side table. I was shown pictures of the same skull with a tape measure on the side. From the jaw to the crown of the head it measured forty-eight centimeters. Not bad. This was followed by two crisp shots of infant skulls, about two to four months old. Pablo pointed to the enlargements of the heads.

"Do you notice anything?" Slowly, I realized the enormity of his find. The infant skulls did not show fontanelles. One must know: the head of a newborn consists of several cranial bones whose interstices grow together only after weeks. About two and a half months after birth, the adhesion to the back of the head closes. This is called a small, posterior fontanelle. The anterior part, the "great fontanelle," takes about two and a half years to become completely covered. The baby skulls in the pictures did not show fontanelles.

Pablo looked at me triumphantly. "These are not earthlings, Don Erich. They are aliens. But born here on Earth."

I learned that Pablo had secured the site and had not informed local archeologists. Only when the skulls were fully documented would he take official steps. He invited me to Colombia. I should bring cameras, two reputable Swiss journalists, and a Swiss notary. Of course, he would cover all costs. I expressed my sincere thanks for the offer and promised Pablo to think about his invitation. First of all, I would have to consult my schedule. I had many international obligations ahead of me. And that was true.

Back at the hotel bar, my fellow travelers welcomed me with great impatience. My explanation didn't seem important to them. They were all talking incessantly. Understandably, because the hotel manager—or Pablo?—had paid for all their drinks.

A few weeks passed. The Anthropological Institute of the University of Zurich presented an exhibition of deformed skulls. During my visit, I met the director of the institute. We started talking, and completely innocently, I told him about my experience in Bolivia.

"The man's name was Pablo?" he asked in disbelief. He wanted me to wait a moment and returned with a newspaper article. It contained a picture of a Pablo who looked exactly like my Pablo.

Weeks and years later I learned in increments the eerie life story of this Pablo. Pablo Escobar was my informant, and back then, Pablo Escobar was the biggest drug dealer in the world. He earned about US$400 million a month, smuggling tons of cocaine across borders. In 1989, Pablo Escobar ranked seventh on *Forbes* magazine's list of the richest men. His fortune was estimated at US$30 billion. Pablo Escobar was considered a cold-blooded killer. He had had hundreds of enemies and friends unceremoniously murdered. In 1991, he had so successfully bribed members of parliament that the new constitution banned the extradition of Colombian citizens. Then he turned himself in to the authorities on the condition that he would never be deported to another country. He was afraid of prison in the United States. He offered to settle the Colombian government's foreign debt. Not far from his hometown of Envigado, Colombia, he built a private prison with bars, lounges, and swimming pools according to his specifications. The public called the complex "La Catedral." He himself had chosen his guards. Everything went well until 1992. Then, alerted by US drug investigators, the authorities found out that Pablo Escobar had continued his criminal activities from the luxury jail. The government had to act, but Pablo was made aware of it and fled. On December 2, 1993, he was surrounded by a special unit. Pablo's son Juan later spread the rumor that his father had shot himself in view of the overwhelming power.

Two weeks after my return to Switzerland, a Mr. Bischofsberger of the American Express travel agency in the Bahnhofstrasse called me in Zurich and let me know that payment for the flights of all participants for my expedition to Colombia had been guaranteed, in business class. I was asked to please set the dates and submit the names, dates of birth, and passport numbers of my fellow travelers. I did not transmit anything. Even though Pablo Escobar's life story was not finished at the time, I nevertheless did not want to have anything to do with a drug dealer. I refused his invitation and did not answer anymore questions from the travel agency. But I remained very curious about the deformed skulls. I could not get the pictures with the baby skulls without fontanelle out of my mind. After all, for decades I had gathered evidence to prove that our planet Earth had been visited by aliens millennia ago. Indicators weren't the same as evidence, though. Did the explanation rest with the deformed skulls? Did they provide the first, solid proof of visiting ETs? Were ancient alien heads really present on our planet? And could this be clearly demonstrated by modern DNA analysis? In fact, such analyses have been carried out. I'll come back to that. But what are these heads really . . . are they deformed skulls? Where can they be observed? How did they take on their form? How old are they? What does science know about them?

Deformed skulls are a worldwide phenomenon, and the oldest ones are well over 9,000 years old. First of all, there are natural explanations for their existence. The deformation of a head can be initiated by the birth process or occur due to lack of space in the mother's womb. Medical science knows many different causes of skull deformities. They are usually classified under the collective term "syndrome." The word comes from the Greek *syndrome* and means "to come together," "to run together." A syndrome is therefore a mixture of several signs of disease that occur together. For example, skull deformities occur in the following syndromes:

- Treacher Collins syndrome
- Beveridge syndrome
- Adams-Oliver syndrome

- C syndrome
- Sturge-Weber syndrome
- Apert syndrome
- Chromosome 15q tetrasomy syndrome
- Chromosome 4q deletion syndrome
- Triploidy syndrome
- Chromosome 7p duplication syndrome
- Ramban-Hasharon syndrome
- Baker Vinters syndrome
- Kenny-Caffey syndrome types 1 and 2
- Laron syndrome types 1 and 2
- Froster-Iskenius-Waterson syndrome
- Theodor Hertz Goodman syndrome

And so forth. These syndromes, which have been discovered by scientists and many of which therefore bear their names, characterize skull deformities due to defects in the chromosomes. These chromosomes in turn consist of DNA sections and belong to each cell. They pass on the genetic information. Each person has twenty-three pairs of chromosomes, or forty-six individual chromosomes. This genetic information can be altered by chromosome defects, which in some cases can lead to skull deformities—as in, for example, the following syndromes:

- Chromosome 11, deletion 11p
- Chromosome 15q, tetrasomy

- Chromosome 7, terminal 7p
- Chromosome 8, monosomy 8p
- Chromosome 9 inversion

In addition to the medical and genetic reasons for deformed skulls that I have already described, there is also "plagiocephaly," which is an asymmetric flattening of the back of the head, and "brachycephaly," which in medicine is understood as a widened skull. And then there is "scaphocephaly," the elongated, narrow skull with a high forehead.

So, are there no mysteries then about these deformed heads? Can everything be satisfactorily explained, medically and genetically?

The Greek physician Hippocrates (460–370 BC), who is credited with the Hippocratic Oath, a set of ethical practices that today's physicians still swear by, wrote in the fifth century about a people called Makrokephaloi. They would put bandages around the heads of newborns to elongate their bones.[1]

The oldest finds of deformed skulls in Europe stem from Armenia and are dated to the ninth millennium BC.[2] But the cult of elongated heads flourished in ancient Europe over millennia. Deformed skulls were found in Hungarian cemeteries. From an ethnic point of view, the custom is said to have originated from the Huns, the Asian equestrian people who moved west from the Caspian steppes during the fourth and fifth centuries. The name of the king of the Huns, Attila, also called the Death Rider, is well known. But the Huns could not have invented the deformations. The custom is much older than the fourth century. Where did the idea originally come from? Gernot Wagner published his diploma thesis about deformed skulls in Austria and comes to the conclusion that the custom had come into the European regions through the nomadic horsemen of the Huns.[3] But—where did they get the custom?

German researcher and writer Hartwig Hausdorf, author of an outstanding series of books on enigmatic finds, talks in his book *Götterbotschaft in den Genen* (*Message of the Gods in the Genes*) about mass skull deformities in France and even in Germany.[4] In the district town of Straubing (Bavaria), archaeologists—according to Hausdorf—"made a surprising discovery during the excavations in a row graveyard in the late 1970s. In two of the tombs they found skeletons with artificially reshaped skulls that were unusually well preserved." According to Hausdorf, there are now twelve artificially deformed skulls that were found in Bavaria alone. All of them are said to date from 450–550 AD. And "at the beginning of the 1930s, British anthropologist Eric John Dingwall (1890–1936) presented a flawlessly researched study that focused on deformities in European countries."[5] Since then, more than eighty skulls are said to have been found in Germany, fifteen in Switzerland, and thousands in France. Isotope analyses, which measure the proportion of a chemical element in a sample, conducted on German, Hungarian, Austrian, and French skulls, showed that they were not from distant lands. They were local. The same question remained: Why did parents for decades torment their children by deforming their skull bones with bandages and boards, often to adulthood? And why did this occur worldwide?

The Museum of Natural History in Vienna exhibits the *Tower Skull of Mannersdorf*. It once belonged to a twenty-year-old man who lived in the fifth century AD. Of the eighteen deformed skulls found in Austria, fourteen lay along a line that ran east through Vienna. Three others appeared in the Krems-Land district on the Danube and another in the Völkermarkt district in Carinthia. Famous scientist Rudolf Virchow (1821–1902), who worked as a physician at the Charité in Berlin, argued that early humans had carried out their skull deformations due to "practical considerations"—or based on a "beauty ideal" or as a means to "distinguish their race from other races."[6] But did they do that to thousands upon thousands of people, and on a global scale, as is now known? Just in Europe alone, artificially deformed skulls were found first in Hungary, from there across Slovakia and Lower Austria to the Rhone, Lake Geneva, and further west toward Burgundy and Bordeaux, even on small islands like Malta. In Bavaria alone there exist "twenty-one documentations from seven sites."[7] And this is—from a global perspective—just a drop of water on the earth's surface. By now we know about hundreds of thousands of deformed skulls. And now—this is hard to believe!—deformed skulls may even have surfaced in Antarctica. What crazy idea has humanity followed for millennia? And in addition to the medically and genetically explainable skulls, in addition to the deformations caused by humans, are there also cone-shaped skulls that don't fit anywhere into the pattern? How did our ancestors carry out the deformations?

We know this from Peru. The Spanish conquerors again and again met young people with elongated heads. They were shown newborns wearing bandages and strings around their heads. Depending on the age and hardness of the skull bones, boards were found to the right and left of the temples. Even boards holding the forehead and latticed plates made of algarrobo wood were used. There were children's cots with a device including a lever at the head of the bed. At bedtime, the baby's head was forced into the device. However, this seemingly unnatural behavior of our ancestors was not started by any South American peoples. The same was also practiced by the Mangbetu tribe in the northeastern part of the jungle in the province Orientale in what used to be Belgian Congo (today the Democratic Republic of the Congo). How did the information about deformed skulls travel from the heart of Africa to the Pacific coast of Peru? Or was it the other way around? Today, most anthropological museums in South America show deformed skulls. The Ethnographic Museum of the University of Buenos Aires in Argentina has a particularly impressive collection. It presents deformed skulls found from Colombia to Ecuador and Peru all the way down to Chile. There, in the dry desert of Atacama, where on the walls of the hills you can admire the gigantic sketches pointing to the heavens, you come to the village of San Pedro de Atacama with its museum Padre Le Paige. The museum was founded by Belgian missionary Gustave Le Paige (1903–1980). He was convinced of the existence of humanlike aliens who are buried in the soil of our Earth. Stored in the basement of his museum, which is not accessible to the public, are more than one hundred deformed skulls. Before his death, he had confided in Chilean journalist Juan Abarzua that he had found mummies with faces that weren't shaped like those of humans. "People wouldn't believe me if I told them what else I found in the graves."[8]

Spanish priest and subsequent bishop Reginaldo Lizárraga (1545–1615), who lived for many years in Peru and spoke the language of the natives, had met with living "longheads." In his *Descripción y Población de las Indias,* he described people with incredibly tall skulls.[9] Originally, they descended from giants, he wrote.

Spanish conqueror Pedro Cieza de León (1520–1554), author of the *Crónica del Perú,* writes about giants that were said to have landed with ships on Peru's coast.[10] Are they the explanation for the deformed skulls? Did giants ever exist? At least in the historical tradition, they do.

They appear in the Bible (Genesis Chapter 6, verse 4): "In those times, and also later, when sons of God joined the daughters of men who then bore them children, the giants existed on Earth. These are the heroes of primeval times, the ones who became famous."[11] The fifth book of Genesis tells of a sarcophagus in which lay the mortal remains of a giant: "See, his basalt coffin is still in the Ammonite city Rabba; it is nine cubits long and four cubits wide" (Chapter 5, verse 11). As the Hebrew cubit measured 48.4 centimeters, the sarcophagus would have been about five meters long. Commonly known is the battle of David against Goliath, described by the prophet Samuel (1 Samuel Chapter 17, verse 4). The prophet Joshua and the Chronicles, which are part of the Bible, also speak of giants. But the Bible is far from being the only literary resource for giants. In the texts about *Die Sagen der Juden* (*The Legends of the Jews*), one can read, "There were the Emins or Terrible Ones, then the Rephaites or Giants . . ."[12] And in the fourteenth chapter of the book of Enoch, which is one of the apocryphal texts, you find this terse statement: "Why did you act like the humans of Earth and fathered giants?"[13]

The giants also play a role in ancient Greek history with regard to the journey of the Argonauts. It happened that the heroes without worry climbed a mountain to get a better view. There they were attacked by giants: "Their bodies had three pairs of nerved hands like paws. The first pair hangs from the horny shoulders, the second and third pair cling to the hideous hips."[14] The monsters also appear in Homer's *Odyssey*, where the hero fought on the island of the Cyclopes against a giant and destroyed his only eye by plunging a burning stake into it. Today, walls made of gigantic stone blocks are called "Cyclopean walls."[15, 16]

Even older is the Babylonian *Epic of Gilgamesh*, which was excavated from a hill near the village of Kujundschick, the former Nineveh (now Iraq).[17] On the clay tablets, which belonged to the library of the Assyrian king Assurbanipal, one can read how Gilgamesh and his friend Enkidu encountered the fearsome creature Chumbaba. It had "paws like a lion, its body was covered with scales."

The book on the origin of the Eskimos states, as a matter of course, "In those days, the giants lived on Earth."[18]

Nothing different is written in the Ethiopian *Kebra Negast* (*Book of Glory of the Kings*). Chapter 100 reports, "Those daughters of Cain, however, with whom the angels had copulated, became pregnant and died . . . because the children split apart the womb of their mothers. As they grew older, they grew into giants."[19]

I wrote about these and more references from ancient literature in an earlier book.[20] If you believe the statements, then these monsters once actually existed. Even Herodotus, the "father of historiography," reported in his first book of *Histories*, "I wanted to dig a well here in the hall, and during my dig I came upon a coffin that was seven cubits long. And because I thought people were never taller than they are today, I opened it and saw that the corpse was really as long as the coffin. I measured it and filled up the hole again." Facts are hard to come by. Over the past few decades, several scientists have affirmed that they have unearthed giant bones or tools of giants somewhere. But all such finds are controversial and their origins are attributed to gorillas or mutations. As late as 1944, German paleontologists Gustav von Königswald (1902–1982) and Franz Weidenreich (1873–1948) reported on the bones of giants that they had bought in pharmacies in Hong Kong. In 1944, Weidenreich even reported this to the American Ethnological Society.[21]

Did giants even participate in the construction of the Great Pyramid? The question is justified because a few years ago a Swiss hobby researcher in Egypt photographed parts of the hand of a mummified giant. Following is the story.

For decades, Gregor Spörri from Basel had been coming back to the country on the Nile. As far as the construction of the Great Pyramid is concerned, he knows both the classical and the skeptical literature, and he personally researched the pyramid several times. Every visit to Egypt attracts traders who try to sell junk to the tourists. To avoid this, on April 14, 1988, Spörri took a ride to the tiny village called Bir Hooker, somewhere between Cairo and Alexandria, hoping to find a dealer who could offer something other than the usual souvenirs. In a house built with mud bricks, he met an elderly Egyptian named Nagib. Nagib offered no figurines, hieroglyphic representations, or papyrus leaves. Instead, he offered real ancient art. After Spörri had told him about his research on the pyramids, Nagib was ready to show him the closely guarded family treasure. This included the chopped-off finger of a giant, including the fingernail and middle and end joint. Everything was covered by mummified skin. Particularly interesting was a bone protruding from the bottom of the finger. Spörri was a smart man and knew how to distinguish fakes from originals. The giant finger could not be a fake. The production of such an object would have been much too expensive. Bones, tissue, the ruptured skin, the fingernail—it was a hideous relic. Not something for tourists, not something to earn money with. Nagib even showed the astonished visitor an x-ray of the giant finger, which was said to have been taken by a friend in a Cairo hospital. Before Nagib put the unique object away, Gregor Spörri was allowed to take a few pictures of it. Back home in Basel he started to think more about it. The giant finger interested Spörri so much that he began to research and eventually wrote a Paleo-SETI thriller in which the finger plays an important role.[22] Luc Bürgin, who publishes the magazine *Mysteries,* became aware of the curious find. He dedicated several pages to the "Monster Claw of Bir Hooker."[23] In its stretched-out state, the finger is at least 38 centimeters long. Truly a big deal. (Mr. Spörri can be contacted directly via *www.grespo.com.*)

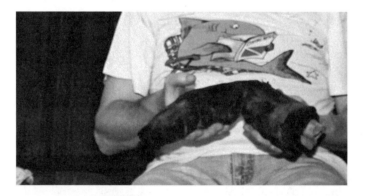

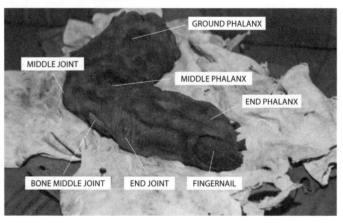

The mummified finger of giant with nail, joints, and even the bone

In a thoroughly researched article in the magazine *Nexus*, British researcher Hugh Newman proves the existence of giants in Ancient Egypt.[24] He refers to Egyptologist Professor Walter Bryan Emery (1903–1971) who was instrumental in carrying out the excavations at Saqqara in the 1930s. Emery found several giant bones and mummies and mentioned them in his excavation reports. These giants are supposed to have been priests and descendants of Horus, according to Emery.[25] They reached a height of up to 4.6 meters. In the forecourt of the temple of Abydos, famous British archaeologist Flinders Petrie (1853–1942) also found bones of beings that seemed to have been much larger than humans.

So do the deformed skulls originate from giants? Something is missing: assuming that there were a million deformed heads worldwide (and in reality, there are more) and further assuming that 80 percent of them were created artificially by man, there are still two hundred thousand "natural" skulls left. About half of them are said to contain genetic factors of giants. But where are the skeletons of these giants? Why are there large numbers of skulls but no bones? The museums of the world exhibit deformed monster heads without skeletons.

In the early 1970s, British zoologist Dr. David Davies discovered an elongated skull in Ecuador (South America) by sheer accident. During a hard rain, Dr. Davies fled into the rooms of a small private collection in the Ecuadorian city of Cuenca. There he found "a great number of fossil material from which two human teeth stared me in the face."[26] Davies inquired about the location of the find and eventually visited it. There, in a rock layer approximately 2 meters high and 2 kilometers long, he discovered bones of various animals that should not exist in South America. Among them were saber-toothed tigers and mastodons (a mammoth species). I was confident of this because I had admired pictures of mammoth-like animals in the collections of Father Carlo Crespi (1891–1982) in Cuenca, Ecuador, and in the collection of Dr. Janvier Cabrera (1924–2001) in the city of Ica, Peru.[27] Among the many different pieces, Davies also came across a deformed skull. He called it "Fred." According to the C-14 analysis, Fred would have to be 28,000 years old and thus the oldest American. Fred's head extended backward; in this case, an obviously natural deformation. But it did not belong to a giant—at least no corresponding bones could be found at the site.

Figurines with skulls extending backward appeared as early as 3000 BC. They are on display at the Museum of Cycladic Art in Athens, Greece.

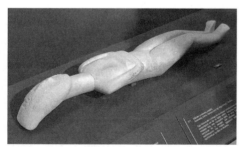

Museum of Cycladic art in Athens—these longheaded figurines and statues are dated back to 3000 BC.

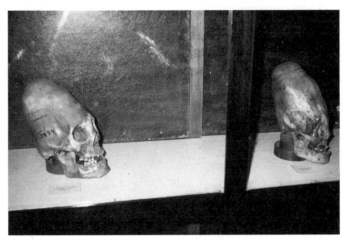

Elongated skulls in the Museo Regional de Ica "Adolfo Bermúdez Jenkins"

"In the Museo Regional de Ica 'Adolfo Bermúdez Jenkins'," writes researcher Reinhard Habeck, "the most extraordinary specimens have survived. Some skulls are so elongated that one does not quite believe their human origin."[28]

And that's exactly it. What perversion caused our ancestors to stretch out the tender heads of their own children? Experts with whom I talked at length did not come up with any sensible explanations. It probably had to do with "practical thinking," such as making it easier to wear headbands. Nonsense; a normal head with a normal forehead carries heavier loads over a headband. Other invented reasons are "beauty ideals" and a way to "distinguish oneself from other social groups." I have a different opinion and explained it in several books.[29, 30]

The worldwide skull deformations were practiced only to make the children look the same as the old gods.

Alien Skulls

All over the world, our ancestors encountered those frightening beings—the aliens. And on a global scale, attention seekers sought to be at least outwardly similar to them. Soon, priests used the barbaric trick of looking godlike through elongating the backs of their heads. It was a great way to impress people. "Look! He looks like a god, and he moves like a god. He must have special wisdom and, thus, a special power over the stupid ones. He must lead them. He deserves respect." The religious leaders wanted to distinguish themselves from the common people by their oversized skulls. The common people needed to see them as the chosen ones. And, of course, the priesthood was owed offerings. For the priestly families, this belief was vital and at the same time hypocritical. They also had to know that the minds of adolescents, despite the deformed skulls, did not function better than the ones of people without elongated heads. In addition, some people with "tower heads" may have been particularly simpleminded. Why? Even though the head was enlarged, *the volume of the brain mass* did not grow with it. That inevitably created hydrocephaly.

My fellow writers Hartwig Hausdorf and bestselling author David Hatcher Childress from the United States arrived at the same conclusion. Hartwig Hausdorf asks, "How would the different peoples on all continents come up with the same crazy idea at the same time? All of them must have met beings whose anatomy was so strikingly different from their own. But since tower heads do not naturally occur on our planet, it seems reasonable to conclude that those beings are not from this world."[31]

David Hatcher Childress, author of over twenty nonfiction books, points to the various theories of skull deformities.[32] There is the "mainstream theory," according to which the heads were altered because of a "beauty ideal." Or the Atlantis theory: the inhabitants of Atlantis are said to have originally possessed elongated heads. People from Atlantis also supposedly visited Mexico, Peru, and Egypt, thereby influencing the local population. The result? Tower heads (the medical term is *turricephaly*). Then there is the "Nephilim-Watchers theory," which holds that at one time "watchmen," extraordinary teachers, would have instructed humankind. These "guards" were said to have lived in the areas of northern Iraq, Kurdistan, and Lebanon and to have visited other peoples elsewhere in airships. The heads of those "guards" were extended backward. So says the theory. And finally there is the "Nephilim Extraterrestrial theory." This one refers to extraterrestrials who had been teaching young mankind.

Do deformed skulls exist today? Whoever sees the solemn entrance of the bishops at the opening of a council in Rome, led by the Holy Father, inevitably associates the bishop's hats, the miter, and papal tiara with the long-drawn heads of former times. The tiara has been the typical headgear of kings since time immemorial. On the mountaintop of Nemrut Dag in Turkey, the Persian ruler Antiochus I wears elongated headgear. The same applies to the Hittite king Tudhaliya IV, represented in the rock sanctuary of Jazilikaja near the Hittite city of Hattusa in Turkey. The motif is ancient. Even the Babylonian master teacher Oannes, who taught man the knowledge of the sciences, is depicted with a raised cap. But he also taught people how to build cities, introduce laws, survey the land, and generally all that is required to meet the daily needs of life—he was a super teacher. And today? After each papal election, the tiara is placed on the head of the Holy Father. This is done with the words: "Receive the threefold crown and never forget that you are the father of the princes and kings, the head of the world . . . ." Whether ecclesiastical or secular rulers, they all copied an ancient custom throughout the millennia. They wanted to look godlike. And this behavior was copied by our ancestors from heavenly teachers, those beings that once came from the stars. We evolved from apes who imitate whatever impresses them using the tools at their disposal. The most incredible impression was definitely left by the gods. Ancient Egypt is a good example of this.

Who was the oldest deity in the ancient land on the Nile, and how was it portrayed? The being's name was Osiris. In the Egyptian imagination, Osiris came from the star constellation Orion. He is shown with the so-called Atef crown, which is nothing but an elongated head. Horus, a son of Osiris, also has a deformed head. Artistic representations of this kind can be admired in every museum of Egyptology. Here is a selection from the Egyptian-Oriental collection of the Kunsthistorisches Museum in Vienna, Austria. Let's look at Amenhotep of the Eighteenth Dynasty, who changed his name to Akhenaten. He and his entire family had long heads. In the Egyptian Museum of Berlin, Germany, hangs a relief that shows him with his wife Nefertiti. The adults play with their three children Meritaton, Maketaton, and

Anchesenpaaton. The children have deformed skulls. So, their heads were not pressed into an artificial form for years. The children were born with deformed skulls! It is common knowledge that the ruling families usually married amongst each other. The logic of this is imperative: the genes of the original gods should be passed on.

Longheaded gods and priests from ancient Egypt on different steles in the museums of Vienna and Berlin

The whole family of Echnaton was longheaded—even the small babies.

And those ETs from far-away planets had long heads too. This can now be scientifically proven. How exactly? Through DNA analysis. DNA is a giant molecule discovered in 1953 by Francis Crick (1916–2004) and James Watson (1928–), who received the Nobel Prize for it. Neither was a biologist. Francis Crick had studied physics and developed naval mines for the British Admiralty during World War II. James Watson was a zoologist and had a hobby studying birds. By sheer coincidence, the two loners came together at Cambridge University and shared their office with each other. In 1951 they began to experiment with DNA.

Already a year earlier, the famous biochemist Erwin Chargaff (1905–2002) had found out that DNA had to consist of four basic building blocks (nucleobases). But how these bases came together and how they passed on their information remained a mystery. Crick and Watson deciphered the confusing code. The giant molecule DNA is comparable to a spiral staircase wound around itself (a *double helix*). You can imagine it as a spiral twisted zipper. Every zipper has teeth (spikes). These correspond to the chemical building blocks guanine, cytosine, thymine, and adenine. They form the "zipper" or DNA strand. It lies at the core of every cell of every living being. If one were to pull apart the DNA strand of a human cell like a thread, it would be almost two meters long. I beg your pardon? How does a 2-meter-long thread fit into a tiny cell visible only under the microscope?

The thread consists of molecular chains, which are atoms that are glued to each other. They are unbelievably tiny and therefore invisible.

And why should one be able to differentiate deformed skulls as being of "earthly" and "non-earthly" origin?

DNA is just as unique as every fingerprint. Although the interlocking molecular chains are similar in all humans, they are not exactly the same. Microscopic bases distinguish one strand of DNA from the other. Every DNA is unique. The building blocks of a person with kidney damage are lined up differently "at the zipper" than in the case of a person with breast cancer. Thus, DNA analysis in criminal science has long become the means to expose the perpetrator or to clear the innocent.

What applies to the individual is also true for all humanity. Today, it is clearly known how the succession of nucleobases of an Asian person differs from that of a European. Although it always remains human DNA—we are all brothers and sisters—we are slightly different from each other. It is undisputed that a Chinese-born person has different facial features than a Native American. To deny this for supposedly racism-related reasons is totally unscientific. The basic characteristics of people in different parts of the world are known. We all remain people, but we are not the same. The genetic patterns are different. No "gender mainstreaming" disputes this scientific fact.

Our geneticists, some call themselves "molecular anthropologists," know, based on DNA, the human development trajectories that made us rise from the hunter-gatherer of antiquity to the urbanist of the present. In every industrialized country today there are companies that conduct DNA analyses of individuals for little money.

A saliva sample is sufficient. In this way, I learned that my ancestors once lived in Mesopotamia between the Euphrates and Tigris.

But time and time again, genetic material emerges on the Earth that doesn't fit into any classifications. Late in 2009, bones were found in the Altai Mountains on the outskirts of Mongolia and Siberia that did not resemble any human being. A thorough DNA analysis by the Max Planck Institute for Evolutionary Anthropology in Leipzig, Germany, revealed: "The distinctly human bones from the Denisova Cave do not match the human genome. The genome of the Denisova man differs from that of Homo sapiens more than twice as much than that of the Neanderthal man."[33] Where did the bones come from? Did they once belong to a deformed skull? More recently, even viruses have appeared on good old planet Earth that can't be classified. They are called tupanviruses—and they do not fit into any modified evolutionary theory. They produce proteins independently and are thus able to create life. So far, a virus has never "lived," because "life," according to the scientific definition, means the organism must be "able to absorb food, excrete, and multiply itself." In order to multiply, viruses have always needed a host. For tupanviruses, the host is superfluous. In addition, this type of virus can repair damage to its own DNA, and that makes it scary. The microscopic monsters were discovered in salt lakes of Brazil and at a depth of 3,000 meters off the Brazilian coast. These are clearly the largest viruses known to date. Ninety percent of their genes are alien and cannot be compared to any other genome, says Patrick Forterre from the Institut Pasteur in Paris.[34] This also applies to some of the long skulls. In the meantime, bones have appeared that cannot be accommodated in the terrestrial genome. How can this be?

The most recent corresponding investigations were suggested by a team led by Dr. L. A. Marzulli.[35] He is the author of twelve books, a radio host, and the lead character in a TV series called *On the Trail of the Nephilim*. Interested in true riddles, his crew also traveled to Peru and filmed deformed skulls in the small Museo Historico Paracas, located on the Pacific Coast. Paracas is known for having burial grounds with so-called Inca mummies, and accordingly, the Museo Historico Paracas also shows deformed skulls in different variants. Initially, the collection originated from the private property of Juan Navarro, a passionate local bone collector. Today, the museum, which in reality consists of only one large room, is managed by Brien Foerster. He published a book about the curious skulls in his collection and thus made the museum well-known.

Mr. Marzulli, Brien Foerster, anthropologist Rick Woodward, medical doctor Michael Alday, chiropractor Dr. Malcolm Warren, and archaeologist Monda Gonzales discussed some of the exciting finds in the Museo Juan Navarro. They then published the results of their research in a book.[36] The most startling result: some of the skulls are from across the ocean and would have come to Peru 500 years before Columbus. Others are not related to the known local DNA.

Alien Skulls

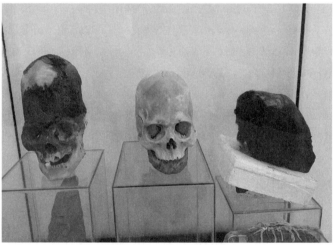

An impressive collection in the Museo Juan Navarro, in Paracas

The team quickly noticed that something was wrong with a number of skulls. Every human head has a foramen magnum—a "big hole" at the base. This foramen magnum connects the cranial cavity with the bony channel of the vertebrae and is therefore the passageway for the central nervous system. Spinal cord and brain converge at this point. In the case of fossil skulls, the location of this occipital hole allows conclusions to be drawn about the posture. The opening in apes is farther back than in Homo sapiens and thus proves the forward bent position of the animal. However, for some of the skulls in the Museo Historica Paracas, the foramen magnum was too far up front—not what one would expect in humans, and any artificial skull deformation did not allow this occipital hole to slip forward. It was coupled with the fibers of the vertebrae at birth. The situation is similar with the occipital condyle. This, too, is closely linked to the bone at the back of the head but is in the wrong place in some of the deformed skulls. The changes cannot be explained by the compression of the skull bones by means of boards, bandages, or strings. They must have been genetically determined before birth.

Consider the Paracas Skull No. 44. It was found in 1927 by the Peruvian father of archeology, Julio Tello (1880–1947), in one of the rock tombs of Paracas. It is said to be over eight thousand years old and is deformed by 25 percent toward the back. On the show *Ancient Aliens* on the History Channel, Brien Foerster, director of the Museo Juan Navarro, said the deformity could not be explained through human influence. The eye sockets are much too large, and the foramen magnum is clearly in the wrong place. As early as 2014, genetic analysis had been carried out on Paracas Skull No. 44. The DNA is not comparable to any local DNA. Meanwhile, the working group around Brien Foerster has had to conduct new analyses. One of them was done at the Paleo-DNA Laboratory of Lakehead University of Ontario, Canada. The result of the scientific research revealed something startling: the DNA of the Paracas skulls was comparable to the DNA of hill tribes in the faraway regions of the Caucasus. The pressing question remained: where did the original Caucasus skulls come from? In addition, the Asian skulls already exhibited the displaced foramen magnum. In his latest publication, Brien Foerster states: "These specific attributes cannot be explained by any cultural skull deformations and they are related to the genetic structure."[37]

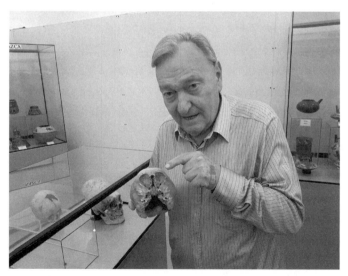

Very strange skull. Seems humanoid, but has a completely different form.

Interestingly, the Museo Juan Navarro also exhibits skulls that have *not* been deformed toward the back and yet they may not have belonged to typical humans either. How so? The skull leaves no room for a brain as we know it. *Our* brain is situated behind the forehead—but some of the skulls have no forehead. The eye sockets are too big for a human and reach down to the nostrils. Why should alien skulls even have DNA? Are ETs not totally different from us humans? Maybe the strangers look like flying elephants or tentacle-swinging bushes? In fact, evolution may have spawned the most fantastic life forms on other planets—but they remain life forms. No matter which corner of the universe they come from, they consist of cells, and each cell is made up of its own DNA. In addition, we should always keep one thought in mind: the gods created the people in *their* own image. How this was possible and why it all came about was discussed in detail in my book *The Gods Never Left Us* (German title: *Botschaften aus dem Jahr 2118*).[38]

Inexplicable skulls exist on Earth. This determination is definitive and confirmed by DNA comparisons. As is known, our brilliant astrophysicists have been searching for traces of extraterrestrial life in space for decades. With forests of antennae, they listen light years out into the universe. So far, we have failed to explore what can be found at our own doorstep. It seemed too "unreasonable." But what is reasonable and what is unreasonable remains a question of the respective zeitgeist. And that changes very fast.

Because skulls exist that seemed to be impossible, the question about the skeleton that carries them remains. The heads did not live "without any support base below." Even extraterrestrial skeletons are now verifiable. More about this later.

CHAPTER 2

Veil Dance around the Pyramids

I TYPE THESE SENTENCES IN THE SUMMER OF 2018. The Greek historian Herodotus (around 485–425 BC) traveled throughout Egypt 2,568 years ago. For months he also stayed in Heliopolis and talked extensively with the local priests and historians. In Chapters 99 and 100 of his second book, he notes: "Everything I've shared so far is based on my own views, my own judgment or my own research. From now on, I want to tell the Egyptian story as I heard it."[1]

We learn that the first Egyptian ruler was Menes, and that he was a god—a "long head." Later, in Chapter 142 of the same book, this is explained more precisely:

> *They recorded in it that 341 generations passed from that first king to the last, the priest of Hephaestus. . . . Now,*

> *300 human generations equal 10,000 years because three generations are 100 years. In addition to 300, one must add the 41 human generations; that makes 1340 years. In other words, in 11,340 years, there has been no god in human form in Egypt.*

Is it all nonsense? Can Herodotus justify his statements? The priests—according to Herodotus—showed him colossal wooden figures. "Because every high priest already has his own statue erected during his own lifetime. The priests counted and showed each one, one after the other, to prove that the son always followed the father. So they went from the image of the last deceased all in sequence to the beginning.... [Chapter 144] They showed that all whose pictures were there, were people of this kind, and that they were very different from the gods. Before these men, however, the gods had reigned in Egypt and lived with the people.... The last of these gods was Horos, the son of Osiris..."

Did Herodotus believe their story of the Egyptians? Definitely yes, because a few sentences later he affirms: "The Egyptians certainly want to know about that because they constantly calculated and wrote down the years of kings and chief priests."

Everything seems pretty confusing, but it becomes even more bizarre. Today, we consider the history of Egypt to begin around 3000 BC with the mystical god kings—not 9,000 years earlier, as Herodotus notes. In addition, Herodotus made his statement about 2,500 years ago. If one adds to the Herodotus years 11,340 years, then this god-king would have ruled Egypt about

13,840 years ago. And Herodotus is very convinced of what he reports. In Chapter 100 of his second book of his *Histories, Das Land Ägypten und seine Geschichte* (*The Land of Egypt and Its History*), he expressly confirms it: "Menes was followed by 330 kings whose names the priests read from a book."

Herodotus visited in the so-called late period in Egypt.

After his visit, only a few pharaohs followed until the dissolution of the classical Pharaonic period. But *before* Herodotus, there were 330 kings "whose names the priests read from a book"? Impossible! Why do these 330 rulers not exist in classical Egyptology? What does today's school know, and what sources are used?

There is the king list of Abydos. The place is located 561 kilometers south of Cairo, directly on the Nile, and thus begin the inconsistencies. The Egyptians of the Old Kingdom—2700–2200 BC—already practiced archeology there. "They ransacked the ground."[2] More recently, in 1726, when no official archeology department existed, the Frenchman Tourtechot Granger (1680–1734) carried out excavations in Abydos. At that time, the entire complex was under the desert sand, only a few upright pillars signaled the presence of any structures below the surface. The next Frenchman to dig in Abydos was Emile Amelineau (1850–1915). In fact, he discovered graves from the First and Second Dynasty—that's about 5,000 years before our present time. The French had settled on Abydos. The next one to clear the stubborn place from its sand was the Frenchman Auguste Mariette (1821–1881) who later founded the Egyptian Museum.

Like his predecessors, he also discovered other graves from the First Dynasty. Finally, by the twentieth century, British explorers Sir Flinders Petrie (1853–1942) and Margaret Murray (1863–1963) conducted research in Abydos. Again, graves from different times came to light. Abydos had once been a very holy place because those who held themselves in high regard wanted to be buried in Abydos. Then—it was now 1912—the famous Swiss Egyptologist Edouard Naville (1844–1926) discovered in the ground under Abydos a gate carved out of granite and several underground chambers. Naville examined the granite boulders of the Osirion, weighing up to one hundred tons, and finally said resignedly that this site must be regarded as "the oldest structure in all of Egypt."[3]

What's is this Osirion supposed to be? I reported about it in my book *Remnants of the Gods* (German title: *Der Mittelmeerraum und seine mysteriöse Vorzeit*) and must recall two abridged sections of it.[4]

Osirion, which incidentally (also spelled Osireion) is a prehistoric megalithic site, located fifteen meters below the level of the temple of Seti I, who built the sites of Abydos. Why is the site called Osirion? Because the head of the heavenly god Osiris is supposed to be hidden there. This at least is what legend tells us. Osiris equals Orion. He originally came from that constellation. But Osiris had a jealous brother, Seth. Seth murdered Osiris and dismembered his body into fourteen parts. The most important part of the body, the head, is supposed to be buried here in Osirion. It was never found.

Veil Dance around the Pyramids

When considering Osiris you inevitably end up at Herodotus' gods who are mentioned in the pyramid texts. The tomb of Teti—who immediately followed the heavenly original god Menes—contains the following writing about Osiris: "Behold, Osiris comes as Orion. Heaven has received you in Orion. You are born with Orion."[5] In the pyramid texts of Unas (2356–2323 BC), Osiris also travels "to the heavenly road." He is a "horizon-dweller who takes off from the Earth in his ship," and who, as statement 303 describes, "ascends to heaven."[6] The statements are unambiguous. Provocatively, I could already ask at this point: What more could you ask for? And this Osirion is, according to Professor Edouard Naville, "the oldest structure in all of Egypt." Most likely he is right. The megaliths that weigh up to one hundred tons are from Aswan, located 400 kilometers up the Nile. The giant blocks used to build Osirion are a perfect fit with Seti I, the builder of the Abydos temple. The stones of the temple of Seti I rest in part on the *megaliths of Osirion*. So Osirion already stood *before* Sethos started his temple construction. It is a distorted world: the oldest technology—the transport and the precise fitting of tons of granite blocks, one with the next, with millimeter precision—is clearly the most perfect one. Seti I, who came along much later, delivered only small stuff. It should be the other way around according to the dictates of technological evolution. *First*, the primitive construction method of Seti I should have been used: smaller stones on top of smaller stones—after all, one keeps learning and gains experience. And only *much later* should we expect the grandiose technology of megalithic art.

Or was the oldest technology the most perfect because the master teachers of young humankind came from the outside; from the skies and from the stars? And in this very temple of Seti I, engraved in a wall, was the famous list of the kings of Abydos. It is a wall with the name-cartouches of the Egyptian pharaohs. On the left side, Pharaoh Sethos I and his son Ramses—the future Ramses II—are recognizable. The engravings show seventy-six rulers of Egypt up to Seti I's time. The following list of names allows a comparison with other royal lists. I am following the spelling used on the German *Wikipedia*.[7]

1. Meni
2. Teti
3. Iteti
4. Ita
5. Sepati
6. Meribape
7. Semsu
8. Qebeh
9. Bedjau
10. Kakau
11. Banetjer
12. Wadjnes
13. Sened

14. Djadjai
15. Nebka
16. Djoser
17. Tati
18. Sedjes
19. Neferkare
20. Snefru (Snofru)
21. Chufu (Cheops)
22. Djedefre
23. Chaifre (Chefren)
24. Menkaure (Mykerinos)
25. Schepseskaf
26. Userkaf
27. Sahure
28. Kakai
29. Raneferef (Neferefre)
30. Niuserre
31. Menkauhor
32. Djedkare
33. Unis (Unas)
34. Teti (Teti II)
35. Userkare
36. Merire (Pepi I)

37. Merenre

38. Neferkare (Pepi II)

39. Merenre II

40. Netjerikare

41. Menkare

42. Neferkare II

43. Neferkare Nebi

44. Djedkare

45. Neferkare Chendu

46. Merenhor

47. Neferkamin

48. Nikare I

49. Neferkare Tereru

50. Neferkahor

51. Neferkare Pepi Seneb

52. Neferkamin Anu

53. Qaikaure

54. Neferkaure

55. Neferkauhor

56. Neferirkare

57. Nebhepetre (Mentuhptep II)

58. Seanchkare (Mentuhotep III

59. Sehetepibre (Amenemhet I)

60. Cheperkare (Sesostris I)

61. Nebukaure (Amenemhet II)

62. Chaicheperre (Sesostris II)

63. Chaikaure (Sesostris III)

64. Nimaatre (Amenemhet III)

65. Maacherure (Amenemhet IV)

66. Nebpehtire

67. Djeserkare (Amenophis I)

68. Aacheperkare (Thutmosis I)

69. Aacheperenre (Thutmosis II)

70. Mencheperre (Thutmosis III)

71. Aacheperure (Amenophis II)

72. Mencheperure (Thutmosis IV)

73. Nebmaatre (Amenophis III)

74. Djesercheperure-Setepenre

75. Menpehtire (Ramses I)

76. Menmaatre (Sethos I)

A total of seventy-six names, starting with the mystical original god Meni (Menes) around 3032 BC and continuing until Menmaatre (Sethos I) of the Nineteenth Dynasty around 1300 BC. Does everything make sense? Where are the missing pharaohs? Finally, Herodotus writes: "Menes was followed by 330 kings whose names the priests read from a book." But the royal list of Abydos has only seventy-six names. To be sure, according to Herodotus, some pharaohs followed until the Twenty-fourth Dynasty; but then Nubians, Assyrians, and Persians ruled over Egypt, and finally, after a "short golden age," Alexander the Great (356–323 BC) invaded Egypt. In 331 BC, the city of Alexandria was founded. (Alexander himself died at thirty-three years of age from an infection in Babylon.)

There is a very long way to go until Herodotus's figure of 330 kings. Egyptology now knows other king lists than just that of Abydos. For example, the royal tablet of Karnak, the Saqqara tablet, or the Turin papyrus. (I'll come back to that.) The annoying thing about Egypt is the fact that no real beginning of its era is known. For comparison, we Christians count the years from the birth of Christ, the Romans counted "ab urbe condita," (after the founding of Rome) in 753 BC. The Maya in Central America counted from the visit of their gods from the universe: 3114 BC. The Jewish calendar begins in 3761 BC—on the day the world was created. But the Egyptians do not know of a fixed date that could be expressed in numbers. It's like sitting in front of a gigantic piece of jelly—there is no fixed point to sink your teeth into. For the chronology following the mystical Meni, alias Menes, the experts painstakingly recalculated their figures from isolated finds with royal cartouches, from buildings and other king lists. This can be a difficult and confusing way, because often pharaohs of the same name occur several times and intersect with other king lists. The resulting data construct is not uncontroversial among Egyptologists, but at least there is a red line one can follow along. Nevertheless, the actual origin, the start date of the Egyptian era, remains unclear. Millennia may be missing. How credible is Herodotus?

The man came from Halicarnassus, a city in the southwest corner of Asia Minor. Herodotus's father rebelled so vehemently against the despot and tyrant Lygdamis of Naxos (around 546 BC) that Herodotus's entire family had to emigrate. On the island of Samos, Herodotus observed the political events of his time. At that time the Persian Empire threatened the Greeks. It was probably the political quarrels that led the young Herodotus to look for first-hand information. He became a "writing globetrotter" of his time and toured Asia Minor, Italy, Sicily, southern Russia, Cyprus, Syria, and even spent a long time in Babylon. In July 448 BC, when he arrived at the place that would later become Alexandria, he began to take notes of what he was told by the people he talked to about their country's past and customs. Herodotus was critical of and often opposed to what he heard "with a certain amount of bias and distrust."[8] Although he grew into the role of a historian, he also reported on the geography and topography of the areas he visited. Herodotus was the first to put the idea to paper that "every story must be viewed in its geographical space, and each geographic space has its own story."[9] At the time of Herodotus there was a lively trade relationship between Greece and Egypt. The Persian king Artaxerxes I (465–424 BC), who then ruled in the country on the Nile, even sent Egyptian boys to Greece for language instruction; conversely, Greek merchants and landlords worked in Egypt. At least at the beginning of his journey, Herodotus did not speak the

Egyptian language. But there were plenty of interpreters for different subject areas, even those who we would today classify as "scientific translators." They provided their services to schools and temples.

Herodotus very quickly differentiated between folk tradition and serious historiography. He could study the latter in numerous libraries in Egypt. When the priests read out the names of the 330 kings to him, he wrote them down exactly, but when he was told a story about a cow, he commented, "I think that's foolish chatter." (Chapter 131) Having become interested, he gathered information about the construction of the Great Pyramid, but complained about details—for example, that during the construction of the pyramid, 1,600 silver talents had been spent for radishes and onions. Herodotus was never gullible; his comments testify to a critical and often sardonic mind. In his reports he neatly separated what was narrative from what was his own opinion. The story of the 330 kings was truly read to him by priests from a book. Herodotus affirms this expressly: "The Egyptians certainly want to know about that because they constantly calculated and wrote down the years of kings and chief priests." (*Histories*, Second book, Chapter 145)

In contrast to the kings list of Abydos, there are names of kings in Herodotus's texts that do not exist on the Abydos list. Examples: "After Proteus, Rhampsinitos became king." (Chapter 121) Cheops, the (alleged) builder of the Great Pyramid, is said to have been the successor of this Rhampsinitos. Herodotus also reports: "After Mykerinos, Asychis ascended the Egyptian royal throne." (Chapter 136) Who is Asychis? Next in Chapter 137: "After him, a blind man from the city of Anysis became king, who was also called Anysis." He was expelled by Ethiopians who, according to Herodotus, ruled Egypt for fifty years. After the (meanwhile very old) Anysis who had returned to Egypt, Seti I became king. But where did Herodotus get the names of kings like Proteus, Rhampsinitos, or Asychis? They do not exist on the king list of Abydos, but they are found in the narrative version of Herodotus's writings.

It's all a bit confusing. In which lost times do Herodotus's missing kings hide? After all, they can only be found in Herodotus's past. Prophetic pharaohs of a distant future were not on any list. But it gets even more complicated. Around 300 BC, a man named Manetho lived in Heliopolis. Commissioned by Ptolemy II (308–246 BC), he wrote three history books on Egypt. Manetho himself was active as a priest. He had free access to all documents in the temple libraries. Manetho's works have been lost, but the Roman-Jewish historian Flavius Josephus (around 37–100 AD), the Christian historian Julius Africanus (160–240 AD) as well as the church prince Eusebius (around 260–339 AD) cited his works. Manetho doesn't just give us seventy-six names as in the Abydos list, but lists even more pharaohs. In connection with the countless kings, Manetho reports again and again about very specific events that are missing from other historians' texts. He seems to have had access to special sources. Manetho deals with vast ancient times that our current Egyptology is not interested in at all.

In volume 1, Manetho reports about dynasties of deities. Eleven of them are listed.[10] Volume 2 lists dynasties twelve to nineteen and volume 3 dynasties twenty to thirty-one. Manetho writes that the first ruler of Egypt was Hephaestus, who also brought the practice of using fire to the people. In Greek mythology, the ruler bears the name Prometheus. Whether in Greece or Egypt, Prometheus was also a descendant of the gods. As in Egypt, he took care of the people: "He took care of his creatures. He taught them to observe the rising and setting of the stars, invented the art of counting for them, the art of writing . . . , but also shipbuilding . . . , and taught them everything else the people needed in life . . ."[11]

In Manetho's list, Hephaestus (in Greek, Prometheus) was followed by Chronos. He, in turn, in Greek mythology, is the father of Zeus. Of course, he came from outer space—where else? Chronos was followed by Osiris (a.k.a., Orion); Tiphon, a brother of Osiris; then came Horos, a son of Osiris; and Isis (a.k.a., Sirius). Manetho states that after Osiris and Isis, the monarchies continued for 13,900 years until King Bidis. "After the gods, the sons of gods ruled for 1,255 years. And other kings ruled 1,817 years. After that 30 memphitic kings ruled for 1,790 years. After that came 10 other Thynitic kings and ruled for 350 years. The kingdom of the sons of gods lasted 5,813 years . . ."[12]

Later, the ecclesiastical prince Eusebius tried to interpret Manetho's numbers—all of them, not just those listed here—as lunar years. But these lunar years, converted into Earth years, still yielded a figure of over 14,000 BC. In our time, the engineer and mathematician Heiner

Paul Kranich dealt extensively with the data of the Egyptian pharaohs. According to his calculations, Manetho's list covers a period of a full 24,925 years.[13] Is Egyptian history much older than we accept? Our diligent Egyptologists dread such thoughts. Every expert knows the famous Sumerian-Babylonian list of kings (now in the British Museum of London), which contains even more impossible numbers. According to this list of kings, the first ten kings ruled 456,000 years from the creation of mankind until the Great Flood. After the Flood "the kingdom descended again from heaven." Again: from where else? The twenty-three rulers *listed by name* after the flood reigned for a total of 24,510 years, three months and three and a half days. Where in the world do we get the precise details of "three months and three and a half days" from?

For comparison: the first ten Babylonian kings in the list just mentioned—ten patriarchs before the Flood in the Old Testament, all had impossibly long lifetimes. Adam is supposed to have aged 930 years, and Seth, Adam's son, was 912 years old when he died. And Methuselah lived 969 years, and so on.

The same is true of Diodorus of Sicily, who worked on his *Universal History* for thirty years (from 60–30 BC). His work encompasses a full forty volumes and begins with the cosmogony, the origin of life in the universe. Of course, the highly educated Diodorus knows all the works of his predecessors, sometimes even mentions them by name and quotes from their works. According to Diodorus, the stars have a connection with mankind. The origin and the father of all beings is a deity. You could hardly express it in a more modern way.[14] According to Diodorus, the Egyptian god Osiris is "a benefactor who visited the whole world." Osiris "taught mankind, he ended its wild behavior and gave man the first civilized form of life." Diodorus asserts that the Chaldeans "watched the stars for many millennia" and taught the eternity of the cosmos. (The Chaldeans were the forerunners of the Babylonians and were admired for their knowledge of astronomy.) But if the Chaldeans observed the stars (for many millennia, according to Diodorus), logically thousands of years of astronomy must have existed. Diodorus notes about the Egyptian era: "Originally, gods and heroes were said to have ruled over Egypt for no less than 18,000 years. The last divine king was Horos, son of Isis. But human kings ruled the land beginning with Moeris for no less than 5,000 years . . ."

Where did the ancient historians take their data from? What Herodotus, Diodorus, and others seem to know is also known to the New World beyond the ocean. The Maya conducted an incredibly precise astronomy, which in turn was only made possible through previous, millennia-old observations, or through a direct training by their heavenly teachers, as some historians note. In the so-called Temple XIX of Palenque, Mexico, a god took over his reign in the year 3309 BC, and on the third tablet of the temple of the inscriptions a date appears in connection with the "boy king Pacal" that lies 1,247,654 years in the past.[15]

In Herodotus's case—and he is the reason for all these cross-connections—what also stunned me was

Veil Dance around the Pyramids

another statement I know no one has stumbled on yet. After reporting about 330 kings and 11,340 years in Chapter 142 of the second book of his *Historien* (*Histories*), which is dedicated to Egypt, he notes: "During this time, four times the sun did not rise in the usual place. Twice it rose from the spot where it presently sets, and twice it set in the east, where it presently rises."

How does a highly educated man like Herodotus write such nonsense? He knows that the sun has been rising daily in the east for ages, traversing the Nile and disappearing to the west. Why should our heavenly star have changed the orbit twice, and above all, how? My astonishment is explained by similar statements of Mayan and several Greek populations who, millennia ago, had no knowledge of Central American cultures. In the *Codex Chimalpopoca*, one of the Old Mexican manuscripts, one reads this: "The second sun was created . . . In it the sky collapsed and then the sun did not follow its path. It's just midday, and then it became night immediately."[16]

This baffling situation becomes understandable when Plato, who lived in Ancient Greece and knew nothing of any Maya in Central America, made similar statements: "Numerous and varied are the devastating ravages that have come upon the human race and will come upon it . . . it is a deviation of the celestial bodies orbiting the sun, and, during long periods of time, an annihilation of the Earth's surface . . ."[17]

Where does Plato's knowledge of the "celestial bodies orbiting the sun" come from? Galileo Galilei was still executed 2,000 years later by the Roman Inquisition because of this same statement. And why did Herodotus, the Maya, and Plato say that the sun was rising in the wrong place?

Have you ever heard of a shift in the Earth's axis? As is well known, the position of the geographic poles of our planet is constantly changing, and at some point the Earth's axis tilts; this is called a *pole shift*. The Earth naturally continues to turn as before, but north suddenly has become south, and the sun in fact suddenly rises in the west. This event is associated with terrible, hard-to-imagine natural disasters.

There is a highly charged cross-link to Fátima, Portugal. There, in 1917, the so-called *Marian apparitions* took place. What would they have to do with Herodotus, the Maya in Central America, and Plato's Ancient Greece? Well, the visionaries of Fátima received three messages from the alleged Virgin Mary, destined for all humanity. The first two concerned the two world wars of the twentieth century. But all popes have refused to publish the *actual* third message. (The cleaned-up version of the third message, released years ago, has nothing to do with its true content. See the following statement by Pope John Paul II, which is in contrast to this diplomatic version.) Was it the knowledge of a pole shift that the popes gained from the apparitions of Fátima but did not dare reveal to humanity? The papal refusal to publish the message was justified by Pope John Paul II on the occasion of a visit to Germany in 1980 as follows: "On the other hand, it should be sufficient for all Christians to know this: if there is a message in which it is written that the oceans will flood whole areas of the earth, and that from one moment to the next millions

Veil Dance around the Pyramids

of people will perish, truly the publication of such a message is no longer something to be so much desired."[18]

These are strange coincidences, aren't there?

But I am still concerned with the grotesque dates of the ancient historians and with Herodotus's 330 kings "whose names the priests read from a book." Besides Herodotus and Manetho, our experts rely on the royal papyrus from Turin. It was discovered in Luxor in 1820 by Italian lawyer and diplomat Bernardino Drovetti (1776–1852) and sold to Turin in 1824. There, in the Museo delle Antichita Egizie, it can still be admired today. When the chest from Egypt was opened, it contained only small fragments of the crumbled original. Who opened the box? It was Jean-François Champollion (1790–1832); he was the only one who could decipher the Egyptian hieroglyphics at that time. How so?

On July 15th, 1799, a French officer, who had traveled to Egypt with Napoleon's troops, at Rashid (Rosette) in the Nile Delta, came upon on a tablet with inscriptions. He didn't know what he had found and presented the tablet to his general who later gave it to the British, and they exhibited it in their new museum in London. There, in the British Museum, the object still attracts thousands of visitors today. On the stone, which was soon generally called the Rosetta Stone, three texts are engraved in various scripts: in Egyptian hieroglyphics, in Demotic script, and in ancient Greek capital letters. The young Frenchman Jean-François Champollion was very interested in the new science of Egyptology. After spending some time in Paris, Champollion was appointed Professor of History at the

University of Grenoble in 1809. Two years later, he returned to Paris, spent hours brooding over the Rosetta Stone, and was the first to realize that all three scripts reproduced the same text in a different language. Champollion could read the Greek. It was a praise from priests of Pharaoh Ptolemy V (204–180 BC). The text begins as follows: "In the year 9 on the 4th of Xanthikos (March 23, 196 BC), . . . of the young man who as King of Upper and Lower Egypt ascended to the seat of his father. He is the living image of Amun and the son of Re, chosen by Ptah . . ."[19]

Starting from the Greek, Champollion deciphered the other two texts and succeeded in creating something like an alphabet of the hieroglyphs. In 1829, the Collège de France set up the first Department of Egyptology in Paris and Jean-François Champollion became chair. In 1824, five years earlier, he had been the one that opened the chest with the Turin Papyrus that had just arrived in Paris from a long journey across the Mediterranean. On the papyrus pieces, Jean-François Champollion recognized some fragments with familiar royal names. After Champollion, the Saxon archaeologist Gustav Seyffarth (1796–1855) examined the fragments and succeeded in putting together several of them in the correct order.[20] The first group of names mentions the dynasties of gods—once more!—who originally ruled over Egypt. Then in the second group, there are thirty "Thinite rulers"—so-called *regents* of the "oldest land." Subsequently, ten "Memphite kings" are listed with names and government dates. From the heavenly god Pharaoh Meni (Menes) to Cleopatra in 30 BC, one hundred sixty-eight kings are verified.

Veil Dance around the Pyramids

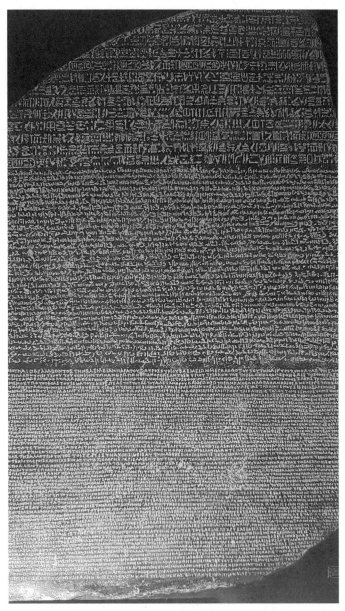

The world-renowned Rosetta Stone. The three different languages on the stone helped to decipher the Egyptian hieroglyphs.

The first four dynasties are listed as follows:

First Dynasty, 2925–2780 BC

1. Menes
2. Hor-Aha
3. Djer
4. Djed
5. Hor Den
6. Anedjib
7. Semerchet
8. Qaa

Second Dynasty, 2780–2686 BC

9. Hetepsechemui
10. Raneb
11. Ninetjer
12. Seth-Peribsen
13. Chasecgemui

Third Dynasty, 2686–2181 BC

14. Sanacht
15. Djoser
16. Sechemhet
17. Chaba
18. Huni

Fourth Dynasty, 2613–2498 BC

19. Snefru
20. Cheops
21. Djedefre
22. Chephren
23. Mykerinos
24. Shepseskaf

A comparison with the kings list of Abydos shows few matches. Snefru, Cheops, Djedefre, Mykerinos, and Shepseskaf are identical. Before and after, however, there are always names that do not exist on the Abydos list. The great experts found ways to mess things up. Very smart Egyptologists sorted and interpreted the lists.[21, 22] This was a difficult task, because names of unknown rulers also exist on clay tablets, stones, seals, or even in graves. For example, the name Scorpion II is on a tomb in Abydos. In addition, the kings were given various titles. But either way, none of the lists of kings arrives at Herodotus's 330 rulers who "priests read from a book." It should be noted here that all the pharaohs who ruled according to Herodotus should not be counted. Artaxerxes I (465–424 BC), Xerxes II (424–423 BC), Artaxerxes III (341–338 BC), or even Darius III (336–332 BC) and the more than 30 rulers of the Ptolemaic period (323–30 BC) have nothing to do with Herodotus's 330 kings. But the priests in Heliopolis did certainly read *330 names from a book* to Herodotus. Neither Herodotus nor the priests cheated. During which periods did the missing pharaohs rule?

There are two more tables of kings created by Karnak and Saqqara. The Karnak list is about rulers who Pharaoh Thutmose III (1479–1425 BC) found represented on a wall when rebuilding his temple. They are illustrations of characters, listed in chronological order. Two new names are added from the so-called "first interim period" that do not appear elsewhere. The Saqqara list was discovered in 1861 in the tomb of a royal scribe (today it can be admired in the Egyptian Museum in Cairo). With a few exceptions, it is identical to the Turin Papyrus.

Like other curious onlookers, I rummaged for days through the different lists of kings, studied intensely the lists of names of the rulers of Upper and Lower Egypt, compared, crossed out, and added and never came up with the 330 kings who were supposed to have governed before Herodotus's time. However, I do not doubt Herodotus's statements. When comparing Herodotus's list with the other lists, it becomes clear that approximately one hundred kings are missing, or about 3,000 years with an assumed 30 years for the reign for each ruler.

But *after* Herodotus, Diodorus of Sicily also reported on Egypt, and this Diodorus was not just anybody. His *Universal History* (Latin: *Diodori Siculi Bibliotheca historica*) comprises forty volumes, the content of which begins with the cosmos and the earliest traditions of mankind and ends at the time of Gaius Julius Caesar (100–44 BC). Diodorus was wealthy and he could afford to travel throughout parts of Asia Minor and study in the well-stocked libraries of Rome and Alexandria. His historical overview covers about 1,000 years. Actually, Diodorus should have

Veil Dance around the Pyramids

known Herodotus's list of the 330 kings. But he did not. Diodorus questions his informants—today we would call them scientists—about the builder of the Great Pyramid. The Egyptian scholars 2,000 years ago should really have known that person. However, incredibly, the "Ancient Egyptians" knew nothing either about the builder of the Great Pyramid or about the time of its construction. Diodorus writes:

> *The eighth king was Chemmis from Memphis. He ruled 50 years and built the largest of the three pyramids, which are considered as one of the Seven Wonders of the World . . . It consists entirely of hard stone, which is very difficult to work with but also lasts forever. No less than a thousand years, it is said, have since gone by to our present day, or, as some write, even more than three thousand or four thousand years, and yet the stones still remain just as they were put together and have kept the entire structure undamaged . . . And the most wonderful thing is that, although works of such sheer size were built and the surrounding area consists only of sand, there is no trace of a dam nor the hewing of the stones, giving the impression that the work was not gradually erected by human hands, but suddenly put into the desert as if by a god . . .*[23]

One has to imagine this: 2,000 years ago, the highly educated Diodorus talked to priests and scientists from Egypt. They had all the documents in their history at their disposal, but yet they did not know how old a gigantic structure in their country was? "*no less than a thousand years, as they say . . . but as some write, even more than three thousand or four thousand . . .*"

That is simply unbelievable!

Diodorus's testimony is underpinned by the greatest mocker among ancient historians, Gaius Plinius Secundus (61–115 AD). Plinius Secundus, also known as Plinius the Younger, born in Novum Comum (now Como on Lake Como), worked as a lawyer, judge, and senator and was one of the universal scholars. He even thought about the spherical shape of the Earth. The family of Gaius Plinius Secundus belonged to the wealthy of their time. Altogether he wrote thirty-one books—one knows that from the writings of his nephew—but until today, only *Naturgeschichte* (*Natural History*) is preserved. His descendants described Gaius Plinius Secundus as a veritable workoholic who read or had people read to him at every opportunity. In his works, he cites his sources and in a preface and even notes that he has read a total of over two thousand books. Its table of contents contains no fewer than 473 headings. In the thirty-sixth book (Chapter 17) of *Natural History* he lists all historians who had already written about the pyramids before him: "The following authors have described the pyramids: Herodotus, Euhemerus, Duris of Samos, Aristagoras, Dionysius, Artemidorus, Alexander Polyhistor, Butoridas, Antihisthenes, Demetrius, Demoteles, Apion. *None of them, however, knows who their true builders are,* and so the creators of this vanity have rightly disappeared into oblivion . . ."[24]

These are twelve historians who already talked about the pyramids before Gaius Plinius Secundus. Most of these men lived in Egypt and used the rich libraries, but no one knew the builder of the Great Pyramid, let alone the year or years during which the structure was

Veil Dance around the Pyramids

built. Among these twelve people, Herodotus is the only one who brought the name Chufu—the Greek word for Cheops—into play, but he explicitly stated that he had been given this name. Gaius Plinius Secundus, on the other hand, *knew* Herodotus's works and commented on the naming of Chufu/Cheops: Herodotus is mistaken.

How can a phenomenal pyramid stand for thousands of years, clearly visible to everyone, admired by the locals and the daily stream of visitors, where visitors must have inquired about the builder all the time, but the Egyptians know nothing about it? The Egyptian guides and scholars must have been bombarded with questions about the pyramid every day, year after year, from sunrise to sunset, for millennia. But they don't know anything about the biggest showpiece of the country? Their knowledge of it has vanished just like Herodotus's 330 pharaohs read to him by priests from a book?

Remember the sun that suddenly rose twice in the west—the pole shift that was associated with terrible natural disasters? At this time, what was once land became water—and vice versa. All coastal towns and those on the plains disappeared, and with them all the libraries. Only small populations in the high mountains survived, and they were busy. Their lives were about survival, food, clothing, and offspring. The second generation after the disaster didn't know anything about libraries from their own experience. After the waters had retreated, the pyramids were still standing. Nobody knew the builders anymore and nobody knew how they had been built. It makes sense when Diodorus notes that, although the area around the pyramid consists of sand, "there was no trace of a dam nor the hewing of the stones." This was logical of course—everything had been washed away. It was even less likely that anyone suspected the real reason for the massive structure, or what was kept in it. What does contemporary research have to say about that?

On April 15, 1993, Ms. Christel Egorov, then spokeswoman for the German Archaeological Institute in Cairo (DAI), said it was known that the Great Pyramid no longer contained any more chambers.[25] And the great, old man of Egyptology, Professor Rainer Stadelmann, added: "It is well known that every treasure in the pyramid has been robbed long ago."[26] The same twisted idea can be heard through all of Egyptology. "There's no way that there are hidden chambers" in the Great Pyramid, assures Egyptologist Aidan Dodson of the University of Bristol in England.[27] This is repeated even though, over the past decades, chambers and shafts have repeatedly been located in the Great Pyramid. The latest results were published by the science journal *Nature* in November 2017. An international physics team had measured so-called muons in the Great Pyramid. These are cosmic particles, 10,000 of which are hitting every square meter of Earth every minute. Completely harmless to humans, these muons move a fraction of a billionth of a second faster as they traverse cavities, and slower as they race through granite. These tiny differences can be measured by modern physics. Thus, in the Great Pyramid, a space of at least 30 meters in length was located, which lies directly above the so-called Great Gallery. The first results have been confirmed by two other groups of physicists with different measuring devices. One Japanese group worked under physics professor Dr. Kunihiro Morishima from the Nagoya University of Tokyo, the other under the direction of the French Atomic Energy Commission in Paris, France. No sooner had the scientists published their definitive results than the archaeologists of the Ancient History Administration in Cairo immediately claimed that the physicists should not have published their results without prior consultation. And why not? Egyptology hates secrets.

It doesn't tolerate any other opinions. A veil of untruth is woven over past and present measurements. Now, you can find the partly secret results of the measurements at the Giza plateau concerning its subsoil and the structures standing on it, which were obtained in the past decades. A thriller of a scandal.[28]

Many scans of the great pyramid in Giza showed hidden chambers and shafts. Illustration by Selina Rueegg.

In 1968, a "spark chamber" was set up in the Chefren Pyramid to carry out radiation experiments. Chief of the project was Dr. Luis Alvarez, who won the Nobel Prize in Physics. At that time, together with scientists from the Cairo Ain Shams University and IBM, more than 2.5 million cosmic rays were recorded. Asked about the results, Dr. Amr Gohed, the Egyptian leader of the team, told journalists that the result was "scientifically impossible." "There is a mystery here that simply has no explanation—call it what you will: occult powers, curse of the pharaohs, witchcraft, or sorcery. There is some force within the pyramids that defies the laws of applied science."[29]

Ten years later, in the summer of 1978, scientists from Stanford University, California, conducted geological drilling around the Sphinx. Although several cavities under the sphinx had been measured, the university did not publish a scientific report.[30] Further explorations took place in the fall of 1980. The Egyptian Ministry of Irrigation used deep drills northeast of the Sphinx. After eighteen meters of debris, they hit a plate of red granite. But granite is not found in the region of Giza.[31] Someone must have deposited the plate many thousands of years ago. No further research was conducted. In the summer of 1986, the two French architects Jean-Patrice Dormion and Gilles Goidin were allowed to perform some harmless measurements with electronic devices in the Great Pyramid. They discovered four cavities; two of them next to the corridor leading to the Queen's Chamber, one in the northeast corner of the Queen's Chamber, and another

Veil Dance around the Pyramids

next to the Grand Gallery. Dormion and Goidin then received permission to drill the walls. After drilling 2.65 meters into the west wall leading to the Queen's Chamber, the drills hit a cavity approximately 3 meters deep filled with quartz sand.[32, 33] Curiously, the sand did not come from the Giza area. Why did the builders of the pyramid fill empty spaces with sand from distant regions? The two Frenchmen were not permitted to continue their explorations.

The next round went to the Japanese. A high-caliber physics team from Waseda University in Tokyo conducted research in the Great Pyramid using a novel radar device that could illuminate different rock strata. In the spring of 1987, the scientists located several rooms and shafts within the structure. The sixty-page, highly illustrated scientific report of Waseda University shows measurement data of the various sections, all of which are traversed with white beams.[34]

One of the passages leads away from the northwest wall of the Queen's Chamber, and a larger cavity was also targeted southwest of the same chamber. Finally, the measurements even showed a 42-meter-long pit that extends below the pyramid. With the report from Waseda University, the antiquities administration in Cairo possessed exact data. Some of the adjacent spaces could have been opened, but nothing happened. At least nothing that was made public. However, the fact that secret drillings were conducted in the Great Pyramid was verified by Alireza Zarei in his book *Die verletzte Pyramide* (*The Damaged Pyramid*) with impressive pictures.[35]

```
┌─────────────────────────────────────┐
│                                     │
│     Studies in Egyptian Culture     │
│               No. 6                 │
│                                     │
│                                     │
│   Non-Destructive Pyramid Investigation (1)  │
│     —By Electromagnetic Wave Method—         │
│                                     │
│                                     │
│                 by                  │
│          Sakuji YOSHIMURA           │
│          (Egyptian Archaeology)     │
│          Takeshi NAKAGAWA           │
│          (Architecture, Dr.)        │
│          Shioji TONOUCHI            │
│          (Geophigics, Dr.)          │
│          Kazuaki SEKI               │
│          (Egyptian Architecture)    │
│                                     │
│                                     │
│                 │
│                                     │
│                1987                 │
│          WASEDA UNIVERSITY          │
│             TOKYO-JAPAN             │
│                                     │
└─────────────────────────────────────┘
```

Title of the Japanese WASEDA-Report from 1987

In the fall of 1992, geologist Dr. Robert Schoch from Boston University (College of Basic Studies), along with other scientists, carried out geological measurements around the Sphinx. Not only were two rooms located under the structure, but the sphinx was also redated.[36] It had to be at least 5,000 years older than previously thought. (According to popular opinion, Pharaoh Chefren, 2520–2494 BC, built the Sphinx.) What was the official reaction to Schoch's measurements? "American hallucinations," said chief archaeologist Dr. Zahi Hawass. And "Schoch's interpretations are not at all based on scientific requirements."[37]

Next it was German explorer Rudolf Gantenbrink's turn. He is an engineer by profession and designed a small robot. The aluminum machine weighed 6 kilograms and was 37 inches long. The technical miracle was powered by seven independent electric motors. Starting from the south side of the Queen's chamber, Gantenbrink discovered a 62-meter-long shaft that ran diagonally upward and ended in front of a small door with two metal fittings. The robot's laser went under the door. There could be a room behind it. Although the discovery was verified on March 22, 1993, it was refuted with lies. First, the German Archaeological Institute in Cairo (DAI) rejected the news: "This is complete nonsense," said the spokeswoman for the institute, Christel Egorov, to Reuters.[38] And: "The discovered tunnel was merely an air duct, and the mini-robot had only been used to measure the humidity in it. It is known that the Great Pyramid contains no more chambers."

But the archaeologists of the DAI knew that their statements were wrong. Gantenbrink's robot had no device for measuring humidity. Professor Rainer Stadelmann, director of the DAI, flatly denied the possibility that there was a chamber behind the small door. "It is common knowledge that every treasure in the pyramid has long since been stolen."[39] His colleague, Egyptologist Dr. Günter Dreyer, confirmed: "There is nothing behind the door. It's all imagination." In the meantime, a small room was photographed behind the door of the Gantenbrink shaft. An endoscope that was passed under the door took several pictures. I reported about it in 2012.

And now the latest news. On November 2, 2017, the scientific journal *Nature* reported: "Cosmic-ray particles reveal secret chamber in Egypt's Great Pyramid."[40] The response from the noble archeology institute of Cairo? "A cavity does not make a chamber," says Zahi Hawass.[41] He added that the Great Pyramid was full of cavities.[42] What a change in the statements compared to earlier ones! And Professor of Egyptology at the University of Basel, Dr. Susanne Bickel, said that if the measurements were correct, the cavity was probably designed to relieve the Great Gallery of the weight of the overlying stone masses.[43] She does not believe that it "had to do with anything mysterious or special burial rites."

Egyptologists hate secrets. They believe everything is scientifically secured and clarified, even if, on closer inspection, nothing is clear. If there are no mysteries in the Great Pyramid, why then are these envied Egyptologists not drilling into the already well-known rooms? And why not do it while TV cameras are watching, sending live footage to the public?

The situation surrounding the Great Pyramid began with Cheops, who doesn't necessarily have to be the builder of this huge stone structure. All the ancient historians who visited Egypt about 2,000 years ago asked the priests about the builder. As I already mentioned, nobody knew the answer. The ancient Egyptian sages did not know who had built the pyramid because this had

Veil Dance around the Pyramids

happened before the Great Flood. Where does the connection between the Great Pyramid and Cheops come from? We owe it to Herodotus. In the second book of his *Historien* (*Histories*), he noted that he had been told that a tyrant named Chufu (the ancient Egyptian name for Cheops) had the pyramid built over twenty years. Now, finally, there was the master who presided over this gigantic project—and everyone would believe it for all eternity, despite the fact that other historians contradicted their colleague Herodotus 2,000 years ago. Even if one were to trust Herodotus, one important question remains. Let's recap: from discussions with Egyptians at the tea table, Herodotus had gathered that Chufu (a.k.a., Cheops) was a tyrant, and over 100,000 slaves had worked on the Great Pyramid for a period of twenty years. It is in the nature of the role: all tyrants are vain. If this tyrant Cheops actually had the pyramid built, it would have to be full of name-cartouches and praises of him. But the Great Pyramid is a monument of total anonymity. Neither in the Great Hall nor in any other room or corridor is there even the smallest inscription. How could a pharaoh build the mightiest structure on Earth without perpetuating his name? The lack of inscriptions is striking. It does not fit the character of the builder.

But hold on a second! Located above the royal chamber were the so-called *relieving chambers*. These are five superimposed cavities above the ceiling of the royal chamber. And there, in the Campbell's Chamber, the word "Chufu" was painted in red. A clear proof that this belonged to Cheops? Not at all. This can be explained by starting with a self-centered deceiver named Howard Vyse. He was a colonel, a perpetual grunt, and a grandson of the Earl of Stafford. Vyse (1784–1853) came to Egypt on December 29, 1835. He had received his excavation license from the British consul. At that time the Italian captain Giovanni Battista Caviglia (1770–1845) was also digging on the pyramid plateau. Initially, they teamed up, but they were so different that they had a falling out, and on February 13, 1837, they ended their friendship. Colonel Vyse chased the Italian from the excavation field.

Seventy-two years before Howard Vyse, the English diplomat Nathaniel Davison (1736–1809) had already discovered a hole in the ceiling at the end of the Great Gallery, which he crawled into on July 8, 1765. He reached an interspace directly above the so-called King's Chamber. Of course, Howard Vyse knew about this "Davison Chamber," because he wrote in his diary that he suspected more voids were above it. On January 27, 1837, he confided to his diary that he *must* discover something before returning to England. Vyse and his engineer John S. Perring got some gunpowder. At the side of the Davison chamber, the men blew a shaft into the strata of rock. On March 30, April 27, May 6, and May 27, 1837, they actually discovered four more cavities above the Davison Chamber. In order of appearance, they were named "Wellington," "Nelson," "Arbuthnot," and "Campbell's Chamber." On the ceiling, Vyse, allegedly by his own admission, discovered some cartouches that might have been smeared on by someone during the construction period. Someone had written, with red brush strokes, the word *Chufu*, but not without errors. This seemed to prove the originator of the structure because Chufu is the Greek version of the Egyptian Cheops. Nobody questioned why a builder was able to write millennia ago, had brush and paint at hand, and painted the Egyptian word Chufu in (false) hieroglyphs.

Now, we know better. British researcher Scott Creighton was able to prove unequivocally that Colonel Howard Vyse, who claimed to have discovered the red inscription, had applied it himself. It was written in 1835 with hieroglyphs that did not exist at Chufu's (Cheops) times. Scott Creighton formulated it in the journal *Nexus*: "There's clear evidence—it's from Vyse himself and undoubtedly confirms he lied about the Cheops Pyramid."[44]

Just a moment. Didn't Dr Bickel, the professor of Egyptology at the University of Basel, say that the newly discovered space in the pyramid serves to "relieve the Great Gallery of the weight of the overlying stones"? So was there a very simple explanation for the cavities in the building?

Sixty years ago, Egyptologist Dr. Hermann Kees already pointed to the nonsense of the relieving chambers by saying: "very original those relieving chambers albeit statically purposeless . . ."[45] I do not know who was first to come up with the unfortunate idea of relieving chambers and why hosts of authors and Egyptologists pass this nonsense on. The relieving chambers are not at all under the top of the pyramid, and they relieve nothing at all. Furthermore, a more than 30-meter-long relieving chamber *above* the already known Great Gallery would make every structural engineer despair. Across from such a relieving chamber, there would have to be a second chamber, just as large, so that the balance is restored and the whole structure is kept stable. In addition, it is thought that the engineers working millennia ago had, while investigating the relieving chambers allegedly planned by the builders, also calculated the total weight of the pyramid. Incomprehensible! What purpose would that have served? The experts do not realize that with every discovery inside the pyramid, the structure looks less and less as as if it had been built by Cheops. After all, every room, every corridor, and every shaft had to be planned before the first stone layer was carted to the site. This means calculations, drawings, the knowledge of the respective material thickness of the different stones, the organization and supply of rose granite from Aswan, and, and, and. Each relieving chamber had to be planned like any other cavity in the pyramid before construction began. After the pyramid had been erected, no 62-meter-long Gantenbrink shaft could be drilled diagonally upward from the Queen's Chamber. Why not? Because the side of the shaft is no longer than 14 centimeters. You cannot even fit a toddler through it armed with hammer and chisel. The construction of the Great Pyramid does not fit with Egyptian society of the Fourth Dynasty. In addition, as Egyptologist Dr. Eva Eggebrecht calculated, the Fourth Dynasty would have had to process a stone volume totaling 8,974,000 cubic meters within the approximately 80 years of reign of the following pharaohs.[46] This for the pyramids of Snefru (2575–2551 BC), Cheops (2551–2528 BC), Djedefre (2528–2520 BC) and Chefren (2520–2494 BC). These four pharoahs form the major figures of the Fourth Dynasty. The daily output would have been 413 blocks. Forget it!

It is time to mention Arab historian Ahmed Al-Makrizi (a.k.a., Ahmed ibn Ali al-Maqrizi, 1364–1442), who, in his book *Hitat* (also written as *Chitat*), reports that the Great Pyramid was built by a ruler named Saurid, and that Saurid was "whom the Hebrews called Enoch."[47] I reported about this Enoch at length in my book *History Is Wrong* [German title: *Falsch Informiert!*][48] For background, in the Bible, Enoch is the seventh patriarch before the Great Flood, and he is the first man to leave Earth in a fiery chariot. Also, the fourteenth-century Arab explorer Ibn Battuta asserted that Enoch had built the Great Pyramid and the other pyramids before the Flood, "to keep in them books of science and knowledge and other precious objects."[49] Hitat reports on treasure chambers lying in the Great Pyramid filled with jewels, implements, "metal that does not rust," glass that folds, "books about the stars," and "treasures of the stars." All of this, along with the complicated design of the pyramid, could well have come from Enoch, who eventually learned the language of the "Guardians of the Sky," spent years in their spaceship, and was taught by them in all areas of science, including astronomy and engineering. According to his own statements, Enoch wrote over one hundred books about his expertise. He is the only one of the ancient authors who even mentions the names of his teachers. Finally, before leaving the Earth in a flying cart, he gave all his works to his son Methuselah with the explicit command to keep them for the "future generations after the flood."[50]

One could actually be satisfied by these statements. The question remains, though, whether Al-Makrizi, the author of the *Hitat*, referred to the right Enoch. Was this the same one who was the first human to leave Earth in a fiery chariot? Or were there other Enochs found in ancient literature? It's about the real Enoch:

> *it is the one whom the Hebrews call Enoch, the son of Jared, the son of Mahalalel, the son of Kenan, the son of Enos, the son of Seth, the son of Adam . . . He had the pyramids built and treasures stored in them "learned scriptures and everything he worried about that it might be lost and disappear, in order to protect and preserve things.*[51]

What treasures are these? What should be kept in the pyramids? The *Hitat* states:

> *All the secret sciences were recorded on the pyramids and their ceilings, walls and columns . . . and the pictures of all the stars painted on them. The names of medical remedies and their uses and harms, as well as the sciences of arithmetic and geometry, and all sciences in general, have been made intelligible to those who know their writing and their language.*

What Al-Makrizi wrote was confirmed more than one thousand years ago by philosopher, geographer, and historian Abu al-Hasan Ali ibn-Husain al-Mas'udi (896–956) in his work *Bis zu den Grenzen der Erde* (*To the Ends of the Earth*).[52] Al-Mas'udi reports that the then-ruling pharaohs were convinced of a coming flood that would destroy their entire civilization. In order to preserve the knowledge of mankind, King Surid Ben Shaluk ordered the construction of an indestructible pyramid. But Surid Ben Shaluk is identical to Saurid from the Hitat and to Enoch from the apocryphal texts. Surid Ben Shaluk had all the valuable books of that time brought to the pyramid, but also seeds of all kinds and the mummified bodies of his ancestors.

All of this suddenly makes sense. The building blocks fit together.

At some point tens of thousands of years ago, people experienced a shift in the Earth's axis. Herodotus, Plato, and the Maya reported just as much about it as did ancient Tibetan scriptures. The entire knowledge was lost. Then the "gods," the masters from outside our solar system, descended. The myths and ancient literature speak a clear language. These "gods" instructed isolated individuals in various sciences. Including a guy named Enoch in astronomy and engineering. They also made it very clear to him that a terrible flood would destroy all of mankind again. Please read more in the book of Enoch.[53]

Enoch is—according to Arabic tradition in the *Hitat*—the same that the Greeks called Hermes. He was their messenger to the gods and was their favorite. Enoch/Hermes felt a great responsibility for the future of the human race. The society of his time revered and admired him. His authority made it possible to gather all the strength and to enforce the construction of the Great Pyramid. The knowledge of humanity, but also the knowledge of the contacts with the "celestials"—the aliens— and their return in the distant future, could never be lost. The "Guardians of the Sky," as he called the teachers, had dictated to him one hundred scientific works. To deal with this flood of information, the aliens had even handed Enoch "a quick-writing device." Enoch's knowledge was intended for mankind of the future—for us. It was never to be lost. I have good reason to believe that the Great Pyramid is a time capsule, created for millennia and indestructible by natural forces. That explains its anonymity. A pharaoh hadn't ordered the building of the pyramid; the gods did. The builder remained Saurid/Enoch/Hermes. Of course, the actual builders were the people of that time. The "gods" never got their hands dirty. The Great Pyramid is a message from the deep past for future generations.

There exist about one hundred theories about *how* the pyramids were built. Most of them have been thought through by engineers and actually sound quite reasonable. But every theory has contradictions, None can conclusively explain the construction process. (I presented ten of the common ideas thirty years ago in my book *Die Augen der Sphinx* [*The Eyes of the Sphinx*].[54]) The fact is, the planning for the myriad manholes, corridors, and spaces within the pyramid is so sophisticated that they can in no way be attributed to Cheops's time. The planning reflects the knowledge of a high-tech society of the past and it seems to have been designed for a high-tech society of the future. The 62-meter-long shaft with its small door at the end, named after Rudolf Gantenbrink, was only discovered due to modern technology. Measuring the new spaces within the pyramid would not have been possible without the capabilities of modern physics. Without our modern technology, no dictator of the past would have had the slightest chance of locating the rooms and corridors in the Great Pyramid, let alone entering them.

In that unknown time before the flood, the next pole shift promptly took place. Enoch had announced it, and Herodotus wrote about it much later. Remember: *"During this time the sun did not rise four times in the usual place. Twice it rose from where it would presently set, and twice it would set in east, where it would presently rise."*

Herodotus's testimony was confirmed in the ancient Mexican *Codex Chimalpopoca*, and the Greek philosopher Plato reiterates his testimony of a terrible catastrophe in several of his works. In the book *Politikos*, wise men of ancient Greece discuss the causes of that annihilation, and Plato diagnoses the cause: "*No, but instead the miracle of the reversal of the setting and rising of the sun and the other stars. Where they now rise, they used to set and they rose from the opposite direction . . .*"[55]

Hesiod lived before Plato and wrote his *Theogonie* (*Theogony*) around 700 BC. Over several pages, it tells of the battles of the gods, which of course were fought in space.

> *the Titans also readied their squadrons . . . , the Earth thundered loudly, the arching sky roared and, immediately, lightning and thunder raced from the sky, holy flames whirled continuously with boulders and flickering lights . . ., the food-sprouting soil burned with loud crashing noise high in the sky, and the mighty forests rattled in embers . . . then the sacred air blazed so that even the eyes of the fittest, stared blinded at the shimmering splendor of thunder and lightning . . . as if the arched sky approached the Earth . . .*[56]

In his famous book *Welten im Zusammenstoß* (*Worlds in Collision*), Immanuel Velikovsky quotes several ancient historians writing about a sun that suddenly rose in the west.[57] (More about that in the next chapter.) And in the tomb of Senenmut (also spelled Senmut, who was one of the architects of the temple of Queen Hatshepsut, 1507–1458 BC), a ceiling painting shows the reversed sky. There, the constellation of Orion suddenly appears west of Sirius rather than east of it. And this with the very exact knowledge of Orion and Sirius that the Ancient Egyptians possessed!

Thus, Egypt was once again under water, unknown thousands of years ago. All libraries were lost and with them all the knowledge. *Therefore*, the Ancient Egyptians did not know the name of the pyramid builder. He had ruled "before the flood."

The same applies to all temples and pyramids of the supposed Fourth Dynasty. The total construction volume of 8,974,000 cubic meters (with a daily output of 413 blocks) was not a creation of the Fourth Dynasty. Everything was already in that location when Cheops entered the stage. It's strange: Snefru, Cheops, Djedefre, and Chefren from the Fourth Dynasty are thought to have processed the 8,974,000 cubic meters of rock during their reign—but none of the builders, architects, engineers, or workers mentioned even one word about the gigantic construction sites—not the smallest inscription anywhere. Why not? Because the Fourth Dynasty had nothing to do with this construction volume. It occurred

Veil Dance around the Pyramids

during the time of Herodotus's 330 kings that the priests read to him from a book.

Any geologist can confirm that the pyramid plateau, along with the (alleged) Pyramid of Cheops, must have been submerged under water for a long time. I wrote about this in an earlier book and included pictures.[58]

The veil dance around the Great Pyramid, the discoveries of spaces that had been suspected for the longest time, the chatter about pseudo relieving chambers and the claims that there are no secrets is a disgrace to our civilization. A group of professionals who are either bona fide naive or do not want to see the long-known facts make fools out of us. Our present technology certainly has the ability to illuminate some of the identified spaces with state-of-the-art geo-radar systems and drill

Geologists clearly prove that Giza with its pyramids and temples was standing underwater for a long time.

into others, then push endoscopes and cameras through the holes to find out what's going on. Obviously, a small group of people want to prevent others from obtaining this knowledge. They call this "responsibility." It is probably not even stupidity that is behind it but simply the naivety of the professionals involved who do not want to see what is quite obvious.

"Nothing but nonsensical conspiracy theories," so mutter the "sensible ones." After all, you cannot deceive hundreds of millions of people. Sorry, but our story proves the opposite. About two billion people call themselves Christians, another billion are Muslims and a few million people belong to the Jewish faith. Each group earnestly invokes so-called "holy scriptures." The fact remains: Christians do not believe in the "scriptures" of Jews or Muslims, the Jews do not believe in those of Muslims or Christians, and Muslims do not believe those of other groups. At least two of the groups with billions or millions of people were misled. It could be that just one group is correct, or none at all. Each group contradicts the other angrily, despite the fact that each of the conflicting groups has brilliant thinkers who warned in their writings and speeches against all the absurdities and subjugations of religions. But few heard the warnings. After all, religious dogmatism remains the main reason for the most terrible wars throughout millennia to the present day. Mankind did not learn anything from it.

That's how we humans are. Once something is stuck in our brain, we defend it, even if it is utter nonsense.

And the assumption that the Fourth Dynasty in the eighty years of the reign of its most important rulers would have processed a total of 8,974,000 cubic meters of rock (including the so-called Cheops Pyramid) is certifiable nonsense.

CHAPTER 3

The Exploded Planet

IN 1949, A BOOK WAS PUBLISHED in New York, and even a quarter of a century later it was still one of the bestsellers in the world: *Worlds in Collision*. The author, Dr. Immanuel Velikovsky (1895–1979), was originally a physician who delved with fervor into ancient philology and ancient history. In Berlin, he founded the scientific journal *Scripta Universitatis atque Bibliothecae Hierosolymitanarum*, to which Albert Einstein (1879–1955) contributed. In 1939, Velikovsky traveled to New York. The plan was to stay for a short while, but the beginning of World War II kept him from returning to Europe. So, he stayed in the United States, moved to Princeton and, until his death, kept a close, friendly relationship with Albert Einstein.

In his brilliant work *Welten im Zusammenstoß* (*Worlds in Collision*) he claimed, "that a planet, namely Venus, used to be a comet and only became a planet with the arrival of human kind . . ."

He hypothesized that the comet Venus had its origin in the planet Jupiter. In support of his theory, Velikovsky cited a wealth of mythologies of various peoples, which in his opinion all pointed to a catastrophe in our solar system. The drama probably played out around 680 BC. At that time, the sun, moon, and stars darkened, the Earth shook and was in disarray, and the poles shifted. "There was more than one world conflagration, but it was worst during the days of the Exodus from Egypt. In hundreds of Bible passages these events are described by the ancient Jews."[1]

The scientific establishment tore up Velikovsky's theories. Neither the astronomers nor the archaeologists were willing to follow him. Exuberant approvals were followed by unobjective and malicious rejections. The sensitive Immanuel Velikovsky suffered under it very much until his death in 1979.

Was he right? Did a catastrophe once occur in our solar system that led to both the Great Flood and the Pole shift? Was that planetary drama the reason for the apocalypses described by Plato, Herodotus, and other ancient historians? I meticulously checked about half of the quotations published by Velikovsky in ancient writings. They all reported a global doomsday. But was the cause actually a comet that people later called "Venus"? Or had Venus long been a planet in 680 BC? Even if that was true, the traditional history wouldn't change. Something terrible must have happened in early human history that affected the whole Earth. Was it a war of the gods? Wouldn't there

The Exploded Planet

be traces of it on our planet as well as in the solar system? And do those elongated skulls that have neither been deformed by humans nor can be explained by mutations come from the survivors of a star war?

Undeniably, all the planets and moons of our solar system are littered with craters. Now, it is generally known that debris from outer space is repeatedly pulled toward planets and strikes their surfaces. Once upon a time, an incredible barrage of boulders must have raced through our solar system. And these would not have been just fragments of mini-meteorites that slowly made their way. At some point, regular missiles must have hit the planets. The impact craters on the moons in our solar system prove it. For example, the craters of Mars's moons Phobos and Deimos have no gravitational power to catch slowly passing meteorites. Phobos has a diameter of 27 kilometers, Deimos is only 15 kilometers in diameter. But both moons are littered with large and small craters. It doesn't matter if these moons have flown into our solar system "from somewhere" because at that "somewhere else," they were not bigger before and their powers of attraction weren't stronger. And between Mars and Jupiter lies the unnatural, planet-free gap we call the asteroid or planetoid belt: hundreds of thousands of larger and smaller boulders are there. How did they get into their orbits?

Fifty years ago, the then leading space scientists still stated in *Weltraumatlas* (*Outer Space Atlas*): "The formation of planetoids is unclear. Either they represent fragments left behind after the formation of the planets, that is, matter that could never combine into a large body because of the destructive influence of Jupiter, or they are remnants of one or more previous planets that existed in the distant past and were torn apart by some fateful process."[2]

Modern astronomy sees it differently: they could be chunks of a planet that was yet to be formed. But the attraction of nearby Jupiter would have prevented this new planet formation. I beg your pardon? If Jupiter's gravitational forces were so strong, then it would have attracted the rocks that were wandering around. And the planetoid belt would not exist because the debris would have been incorporated into Jupiter. Another opinion holds that the total mass of all planetoids is less than that of our moon. That means a planet could not have formed from the existing fragments. What do we actually know about this planetoid belt?

During the night of New Year's Day, 1800, Italian astronomer and monk Giuseppe Piazzi (1746–1826), director of the Palermo observatory, sat behind his telescope working on a new star catalog. A small object suddenly appeared in his field of vision that he had never seen before. He called it "Ceres Ferdinandae." The Roman goddess Ceres is the patron saint of Sicily, and King Ferdinand IV was then the ruler of Italy. Giuseppe Piazzi lost the foreign celestial body, but he had reported his discovery, so other astronomers went in search of it. During the New Year's Eve night of 1801, the German astronomer Heinrich Wilhelm Olbers (1758–1840) rediscovered the object. Today, the asteroid is simply

The Exploded Planet

called Ceres and is around 970 kilometers in diameter, the largest object between Mars and Jupiter. In the following years, further discoveries followed in quick succession. At 560 kilometers in diameter, Pallas became the second largest asteroid, followed by Juno, Vesta, and others. By February 2018, a total of 755,017 asteroids had been registered and cataloged. But there are a variety of asteroids. There are the Amor asteroids. Their paths run between Earth and Mars. The Apohele asteroids have their orbits around Earth. The Atens asteroids orbit the Sun. The Trojan asteroids orbit Jupiter. There are also the Floras, the Nysian, and others. There are even interstellar asteroids that include, for example, Oumuamua, which was discovered in the spring of 2017. At first, our astronomers feared Oumuamua might collide with Earth. The object is quite large—about 400 meters long—and would have caused a global catastrophe upon impact. But exact measurements showed a deviation of the orbit through the gravitation of our sun. Finally, on October 14, 2017, Oumuamua flew past Earth at a distance of about 23,000,000 kilometers. We were quite lucky, human children.

The danger of a collision between Earth and an asteroid still exists. Therefore, the orbits of these celestial bodies are tracked very closely. Our astronomers know that most asteroids from the Aten, Apohele, and Amor groups are near-Earth objects (NEOs). The Jet Propulsion Laboratory (JPL) in Pasadena, California, which is a division of NASA, runs its own program to track and possibly destroy NEOs. The Mars robots were also developed at the JPL and today search for traces of life. Other research institutions, such as the Spacewatch project at the University of Arizona, are also on the hunt for asteroids that could harm mankind.

In the meantime, we know the paths of over 750,000 asteroids. Of them, 158,000 could be dangerous to Earth.

The United States scholar Dr. Nathan Myhrvold, Microsoft's former chief technologist, even claims that the figure of 158,000 is far too low. Countless celestial bodies have been miscalculated and NASA associates misjudged the diameters of many space chunks by a factor of two. In addition, the asteroids themselves are still puzzling. They have different mineral compositions and are in turn littered with craters—where did the chunks that produced them come from?—or they show incomprehensible hues. At the end of June 1997, the United States space probe NEAR photographed asteroid Mathilde from a distance of 1,200 kilometers. Mathilde appeared as a black rock littered with craters. It is believed that the dark coloration is caused by carbon compounds, but surely nobody knows that. Why should only Mathilde be covered with a carbon layer and not all other asteroids? And where do the sixty-seven moons that orbit Jupiter, come from? By the way, among the moon rocks that the United States astronauts brought to Earth, there were also magnetized stones. The moon itself is not magnetic. So where did the magnetic attributes come from? And why should the more than 700,000 bodies of the planetoid belt have in no way come from an exploded planet? Is it because the total mass is smaller than the moon?

The Exploded Planet

Does this argument hold up? After all, a planet does not just consist of a compact mass.

The shell of our earth is very thin. It floats on glowing rock, and in Earth's core, temperatures of around 4,000 degrees Celsius prevail. Nowadays—it's hard to believe!—there are groups that dismiss all these scientifically proven insights as a conspiracy theory and in all seriousness claim that Earth is a disc and that there is no glowing matter in Earth's core. How crazy a society do we live in? The interior of the Earth has been examined several times—ever heard of geothermal energy? And Earth's spherical shape can be proven even without pictures from the moon. (Pictures from Earth's orbit showing the round Earth are even dismissed as fake by some.) What the smart alecs don't know is that at the time of the moon landings, the Americans also installed a small reflective surface on the moon. (Future astronauts will stumble over it; other technical devices also remain on the lunar surface.) As soon as a laser beam fired from the blue planet hits this reflective surface, it is thrown back to Earth in the same hundredth of a second. This makes it possible to measure the exact distance between Earth and the Moon month after month. But now these measurements are made simultaneously from two positions of the globe: from the east and from the west. This results in the ability to measure two angles between Earth and the Moon. If the Earth were a disc, the angles would be different than with a sphere. Easily understood!

Two-thirds of the Earth's surface consists of water, and the continental shelf contains minerals of different density. After an explosion of our planet—whatever the cause could be—the debris on the continental shelf would be thrown out into the solar system; some of it would crash into planets and the Sun, and some would leave the solar system forever. The Earth's interior would become gas, and forced by the gravitational fields within the solar system, some chunks would gather to build a planetoid belt. Due to the existing mass of this belt, future generations could not deduce that there was ever a planet there. Dr. Harry O. Ruppe (1929–2016), for many years professor of space technology at the Technical University of Munich, Germany, considered it entirely possible that the planetoid belt once formed from a planet "destroyed by a catastrophe," and he said, "this planet could have been quite large, if in its destruction the bulk of its matter was hurled out of the solar system."[3]

There are other arguments for a planetary explosion. "The asteroid belt has too much of its own energy."[4] If it consisted of chunks that formed themselves over billions of years from cosmic dust, or if they were meteorite shards that flew in from outside our solar system, then those 700,000 pieces would have other orbits than the current components of the planetoid belt. They would move more sluggishly than they do, and Jupiter would have long ago sucked them into its own gravitational field. The extra energy supports the hypothesis of a planetary explosion. But couldn't "a big comet have collided with a smaller planet?"[5] The computational probability of such a collision is about zero. Therefore, this variant is no longer seriously discussed by experts.

The Exploded Planet

In addition, the existence of an exploded planet can even be verified on Earth. On October 6, 2008, the small asteroid Almahata Sitta crashed into the Nubian Desert in northern Sudan. On behalf of NASA, experts were looking for traces of the celestial body, and they found about 300 small pebbles containing ureilite, a meteorite material. Embedded in this ureilite are tiny diamond fragments, which can only be created at a tremendously high pressure. Experts speak of twenty gigapascals or 200,000 times the air pressure on the Earth. Such energy only comes from inside planets.

Eighty years before Immanuel Velikovsky, a scientist named Dr. A. Frauenholz, at that time geometrician of the Royal Prussian Government in Berlin, proved that once a planet had exploded in our solar system. He published his calculations in the book *Das Sonnensystem in der Vorzeit (The Solar System in Prehistory)*.[6] It reads, *"Due to a catastrophe that occurred in the solar system in prehistoric times, probably at the time of the Great Flood, this planet had dissolved into thousands and thousands of individual parts . . . the long elliptical orbits of most of the asteroids and the comets as well suggest a violent catastrophe, that is to say: explosion of the aforementioned planet . . . And famous astronomers, especially Olbers, have hypothesized that a large planet must have been destroyed by an explosion between the orbits of Mars and Jupiter, from which the asteroids seem to come . . ."*

Even thirty-eight years before Velikovsky, Georg Gerland, who discussed the Great Flood in his book *Der Mythus von der Sintflut (The Myth of the Great Flood)*, commented on the West African tribe of Kanga: "The extermination of the human race did not occur by water but by the collapse of the sky. After all people had been crushed under the rubble of the sky..."[7]

And in 1925, a quarter of a century *before* Velikovsky, the astronomer Dr. Johannes Riem came to the same conclusion.[8] He wrote about Jewish myths that speak of an ancient cosmic event that threatened to destroy Earth. "The Great Flood," says Johannes Riem, "was caused by astronomical constellations." Originally, there was an additional celestial body in our solar system. Dr. Riem refers in this context to the historical accounts of the ancient Persians, according to which "a great, fiery dragon ascended, which devastated everything. The day changed into night, the stars faded, the zodiac was covered by a tremendous tail . . . scalding hot water fell and scorched the trees . . . raindrops the size of human heads fell while the sky was illuminated by frequent lightening . . ." Dr. Riem also reports about the tribe of Kurnai in southeastern Australia. According to their teachings, once a terrible fire fell from the sky. "Then the sea covered the land and drowned nearly all humans. "Far away from Australia, in North America, the Chiglit tribe on the lower Mackenzie River recorded a similar event: "Everything was frightening. The tents of the people disappeared; the wind tore them away. Several boats were tied together side by side. The waves flooded the Rocky Mountain range. The world and the Earth disappeared. Then came a terrible heat . . . The people lamented, they trembled . . ."

The Berjosowo district at the mouth of the Sygva River (tributary of the Ob) in Western Siberia is home to the

The Exploded Planet

Voguls (also called Mansi). They report a "sacred flood of fire that once fell from the sky . . . wherever there was a mountain tree or forest tree, everything was destroyed together with the soil . . ."[9]

According to tradition, the so-called "deluge" must have been closely linked to a cosmic catastrophe. But what caused that destruction? Planets do not simply explode for no reason, and even a real meteor shower originates from somewhere.

On the west side of the Sacramento River lies the settlement area of the Wintun—a Native American tribe previously without contact with the rest of the world. But since time immemorial they report:

> *There once was a world before the one we now live in. This world existed for a long, long time and many peoples lived in it before the present world and we, today's human beings, emerged. In the southern area lived Katichila, a man who, with the help of a magic flint stone, was an exceptional hunter. One day, this ever-present weapon was stolen from him. He told this to his brothers-in-law. To avenge him, these two took off running—the first to the southeast, the second to the southwest—with torches and ran until they reached the point where the sky touched the earth . . . Then they ran around the world and set everything on fire . . . , and all people perished in the fire. The great god Olelbis looked down at the burning world. He could not see anything but flames. The soil burned, the rocks burned, everything burned. Huge clouds of smoke rose up. Fireballs flew from the sky. On his command, a flood began, which rushed over the world in great streams, covering it from the valleys to the mountains, extinguishing the fire.*[10]

We find the same scenario worldwide. Peoples, none of which knew that there were other people living far away from them, recorded the same memories. The book *Matsya Purana*, written in Sanskrit, is part of Indian mythology. But although there were no connections between the ancient Indians and the North American tribes, the essence of their historical narratives in this regard is the same:

> *Then a ring of fire rose up to destroy the heavens, like a set of tongues of the god of death, shining so brightly as if twelve suns would be rising. When the world was scorched by a number of terrible suns, all parts of the world, moving and static, were immediately reduced to ashes. Hereupon a mass of coal-black clouds appeared, reminiscent of the scorched world. Then the weight of violent downpours continued to pound the Earth. As if the regions of the worlds wanted to shed a stream of tears of pain over the destruction of the universe.*[11]

In the Vatican Library of Rome is a manuscript that the Spanish conquerors brought back from the New World. It must have been created over 400 years ago in the highlands of Mexico and was originally written by Mayan priests. The text is labeled *Codex Vatican A 3738*. In it—as in many documents of other peoples—is the account of a destruction of the world by heat, fire, and wind. The sky had glowed, the fire had rained from the sky, and the people had burned. Hot ashes were present everywhere like blistering lava.

Siberia is located far from Central America. A connection between the two cultures did not exist millennia ago. But even the northern Siberian tribes report "a terrible flood of fire that once destroyed all life on Earth."[12] The same applies to the Tungusic people of Transbaikalia. According to their tradition, a heavenly conflagration once raged for seven years and destroyed everything. All but one boy and one girl are said to have died, and they only survived, "because they flew towards the sky on an eagle." The aforementioned Mansi (formerly known as Voguls) in Western Siberia also report that the supreme heavenly god caused the flood of fire in order to destroy the devil that ruled Earth. While the whole Earth drowned in fiery water, the gods had succeeded in saving some people in an iron ship. The inhabitants of Tahiti also report something similar. And as is well-known, Tahiti is located in the Pacific Ocean, tens of thousands of miles away from the Mansi. "Glowing rocks" had fallen from the sky. After the catastrophic fire, "there was no hut, no coconut palms, no fruit, not even more grass."[13] And how do the Ipurina, Pamary, and Abedery tribes that live on the Purus River know the same? The Purus River flows east of the Andes (Brazil), and the jungle tribes of the Amazon could not know about the Mansi in Western Siberia. However, they all know the millennia-old accounts of fire and glowing rocks that crashed from the sky.

And it is always the gods who played the decisive role. Either they had fought a battle in space that led to the destruction of a celestial body, or they had emerged to save at least some people from destruction.

Even in the national epic of the Finns, *Kalevala*, hot rocks are reported to have once fallen from the sky.[14]

Where would rocks fom the sky have come from? The occasional errant small meteorite may crash into the ground, unless it first burns up in the atmosphere, but hundreds of them at the same time? Even the Book of Books, the Bible, describes it: "And when they were at the foot of Beth-Horon, fleeing from Israel, the Lord had large rocks fall from heaven as far as Azekah, and they died. Those who died from the hailstones were more than those who Israel slew with the sword . . ." See Chapter 10, Verse 11, of the book of Joshua.[15]

Anyone who delves into the history of earthquakes quickly learns that hundreds of narratives exist. The topic captivated astronomers, ethnologists, and theologians across human history. What happened unknown millennia ago? How could a worldwide flood occur as reported by all cultures on Earth? How was it possible that the sky was burning and glowing rocks were falling? Nothing comes from nothing—there is a cause behind every event. In the more than one hundred books that I studied about global "floods" and "fires from the skies," the authors always assume that the cause is a "divine judgment." Some heavenly being had been angry, either at men or other gods. But in reality, there is no god, neither in the past nor today, who constantly rages against his own creatures because they do not do what he wants. For the destruction of Earth, whether its caused by a great flood or a catastrophe in the solar system, there must be another cause than a constantly raging deity. What is this anyway—god? The philosophical question has a lot to do with the gods and the annihilation of one planet in our solar system.

The universe began at some point in time. Religion explains this origin with god. But where did this god come from? Who or what created god? Science sees it differently: in the beginning was the Big Bang. But even a "primordial atom" has an origin. Nothing comes of nothing, even in astrophysics. Whether it's religion or science, if we leave speculation aside, we have zero knowledge with regard to the origin of the universe. And whoever answers this question with a "divine being" swiftly drowns in a sea of impossibilities.

With a (supposedly) divine being that created the universe, man thinks as illogically as with regard to the Big Bang. In the realm of faith, a divine being is possible—but a Big Bang out of nothingness isn't. In this case, a "divine being" would have to possess some minimal qualities without which creation is unthinkable—at least for us humans. One of these minimal properties would be timelessness. A real god—a spiritual being who creates something out of nothing, whatever he wants—would not have to wait and see how his experiments turn out. He would know this beforehand, and thus the experiments would be superfluous. But now tens of millions of people believe that an angry god has conjured up the deluge and annihilation of the world through heat. That's how it is told in the holy books. Those satisfied with such answers no longer need research, and those who do not do research cannot find the cause of a deluge or the emergence of the asteroid belt. In this ignorance, man does not know that the catastrophes of the past can be repeated in the future. And the person without that knowledge has no ability to avert a future destruction of the Earth. So, is a god responsible for the annihilations of the past? To focus on this question, let me go back a bit.

In the beginning—as the astrophysicists write—it was nothingness, endless emptiness, or "black radiation." This radiation is said to have existed in a state of "non-time" or "expectation." Whoever asks what existed *before* this state will never find an answer. Then out of the black radiation is said to have emerged a first particle of matter: the electron. This electron combined with a proton—where did that come from?—and a hydrogen atom was formed. Matter was born. Millions of years after the Big Bang, cosmic dust swept through the universe. The particles swirled around and found each other, eventually forming a sphere that attracted more and more matter. *Where* did this matter come from? The growing density caused a friction between the particles. The matter became warmer and finally hot. A red-hot star was born. It condensed even more and attracted more and more cosmic dust. Light atomic nuclei merged into heavier ones. Hydrogen became helium, and later other elements developed.

There is a reason for my excursion into astrophysics. I am tracking this "divine being." The "thing" that is said to be responsible for the catastrophes of the past. That's why I have to follow the creation of the universe a bit further. During the fusion process within a sun, energy is produced and released, and nuclear fusion takes place, which is provable with our Sun. But as soon as the lighter elements are used up, this nuclear fusion ends because there is nothing more to fuse. The star inflates, explodes, and becomes a supernova; thereby a lot of the star's mass is thrown into space. In the final phase of extinction, the greater part of the star mass falls back into the star and compresses. The result is a *white dwarf*. The diameter of this white dwarf shrinks to just a few kilometers. The star is at the end and collapses. After it collapses, not one ray of light speaks to its former existence. What remains is a black hole. Astrophysicist Reinhard Breuer describes it as follows: *"A black hole is a star that has become so heavy through contraction that no particle, not even light, can escape its surface."*[16]

Comparable to a bubble in the water, the black hole forms space within space. What was once trapped in the space of the black hole never comes out again. The scary behemoth doesn't even let light escape, so it is invisible. It reveals its existence only through the curvature of space, which can be demonstrated in astrophysics. In this alien world—and all this has to do with the (assumed) god—the physical laws are completely different from our environment.

- Time runs in reverse in the black hole.

- Space is of *temporal* nature and time of *spatial* nature.

- Within the black hole, all happens with decreasing entropy. This means that the "order" is becoming higher and higher. (In plain English: if you pour a jug of hot water into a bathtub of cold water, the hot and cold water mix, creating "disorder." The hot water does not remain separate, and neither does the cold water.)

- Time runs cyclically in a black hole. All past states are repeated again and again. Every piece of information always returns to its starting point.

It all sounds confusing. And now imagine a "friendly god" who created all this and, for tens of billions of years, has watched how this whole process—the birth and destruction of stars, the emergence of white dwarfs and black holes—repeats itself over and over again. If this god had a personality, he would have been bored out of his mind for ages. And if this friendly god would be immaterial, and purely spiritual, then he would have to be even more bored. Because to know everything about what is happening and what will happen results in an even bleaker existence.

Furthermore, one would realize that the universe consists of an infinite number of suns and planets, including trillions of worlds with life forms, including human-like forms. Thus, according to an extrapolation by NASA, there are 4.5 billion Earth-like planets in our galaxy alone. And somewhere out there would be a friendly god who created all this. This god picks a blue world among the trillions of planets and creates intelligent beings—humans—and lets them develop. And among them, he prefers a special people. He gives his commandments, and because they do not do what he wants, he rages, rants, and threatens—all as recorded in the Bible. Yet, as a timeless being, he must have known from the outset that his creatures are stubborn donkeys, so there is no reason for the anger. But the Bible teaches the opposite. Would you care for some examples?

> *"They should die from epidemics, they should not be mourned and not be buried. They are to become manure in the fields, perished by the sword and by starvation, and their bodies should become food for the birds . . ." (Jeremiah, Chapter 16, Verse 4)*

> *"The Lord said, 'The dogs shall devour Jezebel. . . . Those who die by Ahab in the city shall be eaten by the dogs . . .'" (1 Kings, Chapter 21, Verse 24)*

> *"If you turn away and ignore my statutes and commandments, I will destroy you . . ." (2 Chronicles, Chapter 7, Verse 19)*

> *"They are locked up in the pit, locked up as prisoners, and they are locked in a cell and, after many days, will be punished . . ." (Isaiah Chapter 24, Verse 22)*

> *"Behold, my servants will eat, you will starve. See, my servants will drink, you will be thirsty! Behold, my servants will rejoice, ye shall cry out in heartbreak . . . may the Lord God kill you . . . " (Isaiah Chapter 65, Verse 13)*

> *"But I will chastise you all . . . I will pour out my wrath over them like water . . . I am like a moth and like a worm-eater . . ." (Hosea, Chapter 5, Verse 3 ff.)*

> *"You shall be damned in the city and cursed in the field . . . The Lord will release the curse onto you, the consternation and the threat . . ., the Lord will attach the plague to you until he exterminates you . . ., the Lord will beat you down with consumption, with hot fever, with drought, with the*

burning of crops and yellowing . . ., the Lord will turn the rain of your land into dust, from heaven it will come down over you, until you are consumed. If you do not faithfully fulfill all the words of this law . . . the Lord will torment you and your descendants with selected plagues, with great, ongoing plagues . . . the Lord will take pleasure in annihilating you and destroying you so that you will be torn away . . ."
(5th Book of Moses, Chapter 28, Verse 16 ff.)

This goes on from one Biblical author to the next. Even the absurd becomes possible: "The king desires no other bride's price than a hundred foreskins of Philistines to avenge the king's enemies . . ." (1 Samuel, Chapter 18, Verse 25)

Why these quotes? What do they have to do with the emergence of the asteroid belt or the Flood? The quotes suggest a god who can never be a "grandiose spirit of creation." An angry, vicious god crying for revenge would be an insult to a spiritual being. Like all other gods of mythology, the one of the Old Testament was nothing else but an extraterrestrial. This statement is provable. And in this way of thinking, the real God, that "thing" incomprehensible to man, whom I call the "grandiose spirit of creation," is not lost. For my part, I have long since unmasked the gods of religions and mythologies, and have, nevertheless, certainly never lost faith in a real god. But one cannot attribute the tantrums, threats, and punishments to the true God.

The very last doubt about the alleged god of the ancient scriptures is definitively removed by the so-called prophet Enoch. See his book of the same name.[17] Enoch

The Exploded Planet

reports how he was brought beyond the Earth and further to the "sanctuary" (the command center of the spaceship): "Then the Majesty rose, came to me and greeted me with her voice"; this is embarrassing for theological scholars because this event cannot be rationalized as a friendly god. Enoch tells us how he was picked up by two aliens ("as I've never seen any on Earth"), disinfected ("rubbed with a heavenly ointment"), put into a spacesuit ("See, I looked as one of them"), and was brought to the mother spaceship and to the commander. He welcomes him "with his voice." He gives instructions to hand Enoch a "rapid-writing-tube," and a subordinate named Vrevoel dictates scientific works to Enoch for months. Before his "ascension," Enoch passes these works to his son Methuselah with the words: "And now my son Methuselah, keep the books of your father's hand, and hand them over to the coming generations after the Flood . . ." (Enoch, Chapter 82 ff.) By the way: Enoch's books are kept in the Great Pyramid of Egypt.

So if an almighty God cannot be held responsible for the explosion in the asteroid belt and the Flood, if it was *not* "the grandiose spirit of creation," who is left? Were these all natural events or cosmic catastrophes? As for the Flood, the proof is easy: it was not about a natural event. Why not? Here is a thought from my most recent book:

> *Flood stories are a global fact within religions and myths. Whether in the Bible, in the Sumerian Gilgamesh epic, in the "Song of Creation," the Enuma Elish of the Babylonians, whether in the* Book of Mormon, *the Kogi in Colombia, the Hopi in Arizona, the Maya in Central America, or the*

Dogon in Africa—it is always a "god" warning people about the flood. They all knew about the impending water masses; and not just two weeks in advance, as in today's weather forecasts. The time of the Flood was determined much earlier, the countdown was on, and all the pupils, without exception, were instructed to build a specific watertight ship. Logically, none of the gods seemed "almighty," no one just gave the people a boat. None of the "celestial beings" who had come down had the omnipotence of conjuring up a ship, so to speak, by winking or snapping their fingers. Technology was used in building every ship. Shipbuilding is a technology. And each of the pseudo-gods called the coming Flood a punishment. Mankind of that time was to be destroyed. Obviously, a genetic program went awry. At least that's how it has been handed down to us several times.[18]

"A great disaster will engulf the whole Earth, a deluge . . . the complete weight of judgment will come upon the Earth and the Earth will be cleansed of all filthiness . . . That is why I bring a flood upon the Earth." (Book of Enoch, Chapter 106, Verse 13 ff.)[19]

And the asteroid belt? Was there a war in outer space? Did some pseudo-gods rage in our own solar system and want to destroy each other? The biblical prophet Daniel writes of a heavenly army that was thrown from the stars down to the Earth. (Daniel, Chapter 8, Verse 10) And John the apocalypticist tells us of stars falling from the sky and a war in space. But his visions never make it clear whether he means the past or the future.

"And the third angel sounded his trumpet; Then a great star fell from the sky, burning like a torch, and it fell on the third

part of the rivers and on the springs of water . . . "(Revelation, Chapter 8, Verse 10)

". . . . And the fifth angel sounded his trumpet when I saw a star that fell from heaven to Earth." (Revelation, Chapter 9, Verse 1)

". . . . And there was war in heaven, so that Michael and his angels fought a war with the dragon. And the dragon waged war, and his angels; and they could not withstand, and there was no more place for them in heaven. And the great dragon was thrown, the old serpent, called the devil, and Satan, who seduced the whole world, he was thrown down to Earth, and his angels were cast out with him." (Revelation Chapter 12, Verses 7–9)

Where do these texts come from? The apocalypse is in every Bible in the New Testament appendix. It was said to have been written by the apostle John. But several theological experts believe that the text came from an editorial team from the 90s to the 100s (AD) of our time.[20] But even this editorial team has not invented the content—it used older sources that are no longer available to us.

In previous books, I have referred to the Indian historical accounts, which talk about a war of the gods in our solar system.[21] The following text from Sanskrit dates from 1888 and was translated by Dr. Pratap Chandra Roy (1842–1895) at a time when we humans could not know anything about space cities and wars in the solar system.

The gods who had fled returned . . . originally the valiant Asuras had three cities in the sky. Each of these cities was large and well built . . . (v. 77, 691). Civa was preparing for

> the destruction of the three cities. And Sthanu, the Annihilator, commanded a unique battle position. Then, when the three cities met in the firmament, the god Mahadeva pierced them with his terrible beam from threefold belts. The Danavas were unable to withstand this beam of Yuga fire consisting of Vishnu and Soma. As the three cities began to burn, Parvati rushed there to watch the spectacle.[22]

And does the story of a battle in heaven not also haunt the Christian world view? An archangel named Lucifer appeared there with his legions and came before the throne of Almighty God. "We no longer serve you," he commanded, and the friendly God instructed the Archangel Michael to throw Lucifer and his gang out of the heaven. That "heaven" must definitely have been the universe, because in a "divine" heaven, paradisiacal states would prevail without any struggle.

In the *Mythos der fünf Menschengeschlechter* (*Myth of Five Races of Men*), written around 700 BC by Greek poet Hesiod, a war of the gods is mentioned: "Those heroes of noble race, called demi-gods, who inhabited the infinite Earth in the time before us, were doomed by awful war and terrible battles."

Hesiod also explains who those fighting gods were: "In the beginning, the golden race of mortal people worshipped the immortals who inhabited the Olympic Houses. These are Chronos's comrades, when he still reigned in heaven."[23] A former battle in our solar system haunts the myths. They are not historiography, but folk tradition. Even the Greek legend where the creation of the human race is described tells of a fight between

The Exploded Planet

Zeus and Chronos. And Zeus—the thundering, the thunderbolt—fought against the dragon Typhoeus. In outer space; where else?[24]

I already pointed to Plato, who explains the Phaethon legend in his dialogue *Critias* and explicitly makes clear to his debaters, who were the scholars of their time:

> *". . . but in reality it is a deviation of the celestial bodies circling the Earth and a mass destruction of the Earth's surface that is repeated in long periods of time . . ., that is why you will always become young again, without any knowledge of what happened in ancient times . . . The founding of our state system here, according to the recordings of the temple documents, occurred eight thousand years ago . . ."*[25]

Eight thousand years was a long time ago, but Plato knows what he's talking about. He refers to "temple documents." And according to them, the cause of destruction was said to be "a deviation of the celestial bodies orbiting the Earth." Clearly a war was fought by those "gods" who were constantly raging and used terrible weapons against each other?

Another variant offers the myth of Phaeton. This is not about a war of the gods among themselves, but is the fault of Phaeton—the son of the sun god Helios—who cannot steer the sun chariot and unintentionally causes terrible annihilations. It should always be kept in mind that the Phaeton myth is verifiably thousands of years older than Plato. "The original story does not exist in ancient testimonies of Greek mythology handed down to us. We do not find it in Homer's writings, not with the Greek poets, and not in the work of Hesiod that has been handed down to us."[26]

In 1955, the then most famous sumerologist, James Pritchard (1909–1997), was translating Akkadian myths and epics. Among them is the *Epic of Creation,* a text whose origin cannot be dated any more than other traditions. As in the Bible, there is talk of the battle of the gods against a sky dragon. "Mother Hubur created monster snakes," against which other weapons were ineffective." "Roaring dragons seemed like gods in the sky."[27]

Everyone knows the fire-breathing dragons in the Chinese traditions, and quite a few of them led wars in space. Legendary Emperor Chih Chiang Tzu-Yu (who also appears as Hou Yih) is said to have undertaken flights into space and fought "against ten suns."[28] Two thousand years ago, Gaius Plinius Secundus wrote in his *Naturgeschichte* (*Natural History*): "A formidable comet was observed by the people of Ethiopia and Egypt, and named after the king ruling at this time, Typhon. It had a fiery appearance and was twisted like a spiral, and he was very grim-looking. It was not so much a star as something you can call a fiery orb."[29]

When we talk about battles of the gods in space, *Das Buch des Dzyan* (*The Book of Dzyan*) must definitely be included. The origin of the related doctrine is unknown, but at least we know today that the word *Dzyan* denotes neither a prophet nor any god. Dzyan is rather the name for a summary of Indian and Tibetan histories. These texts were compiled by Mrs. Helena Blavatsky (1831–1891). She was the founder of the Theosophical Society and published *Das Buch des Dzyan* in 1888 in London. In the foreword, the very well-read lady, who was also able to translate scriptures from Sanskrit, assured her readers that all her sources came from ancient Indian and Tibetan libraries. She also published the names and locations of these libraries. Nevertheless, Mrs. Blavatsky was still harshly criticized during her lifetime by German Sanskrit scholar Dr. Max Müller (1823–1900), Professor of Indology at Oxford. However, the most important Sanskrit expert at that time, Dr. Dayanand Saraswati (1824–1883), vehemently defended her and confirmed her sources. Essential parts of the *Das Buch des Dzyan* were incomprehensible one hundred years ago because they dealt with technologies that were only going be realized in the future. Who would possibly understand that?

Imagine huge spaceships in which several generations live. Such structures will be in the form of wheels that continuously rotate around their own axis. Why? Self-rotation

Bigger spaceships will always have a round shape. Through rotation you get gravity.

generates centrifugal force inside the spacecraft, which creates an artificial gravity. The spaceship crew does not float weightlessly—they always have ground under their feet. Then the inhabitants multiply and build spaceships for new generations and fan out into different directions in space.

Mrs. Blavatsky could not have conceived of such a scenario in 1888—but the *Das Buch des Dzyan* describes gigantic space wheels and the increasing population of the inhabitants. And also battles between the gods.

(Stanza 1) ". . . There was no time, for she was asleep in the infinite womb of duration . . . darkness alone filled the infinite universe, for father, mother and son were once again one, and the son had not yet awakened for the new wheel and his journey . . . life pulsated unconsciously in space . . . but where was Dangma when the Alaya of the universe was in Paranartha and the great wheel was Anupadaka?"

(Stanza 2) ". . . where were the builders, the shining sons of the dawning Manvantara? . . . The producers of form out of non-form, the root of the world? . . . The hour had not yet come; the beam had not struck the seed; Matripadma was not yet swollen . . ."

(Stanza 4) ". . . sons, listen to your teachers—the sons of fire. Learn that there is neither first nor last; for all is a single number that has come out of the non-number . . . Hear what we, the descendants of the original seven, born of the primordial flame, have learned from our fathers . . ., from the glory of the light that shone from eternal darkness, the awakened energies sprang up in space . . . "

(Stanza 5) ". . . when he begins his work, he separates the sparks of the lower wheel, which float joyfully in their radiant

dwellings, and builds from these forms the seeds of the wheels. He places them in the six directions of space and one in the middle: the main wheel . . ., a host of the sons of light standing in every corner, and the Lipika in the middle wheel. They say that is good. The first divine world is finished . . . Fohat takes five steps and forms a winged wheel in every corner of the square . . . "

(Stanza 6) ". . . finally, seven small wheels are turning, whereby each gives birth to the next . . . he builds them as images of older wheels and affixes them to imperishable centers . . . how are they built by Fohat? He collects the fiery dust. He makes orbs of fire, runs through them and around them, and gives them life. Then he sets them in motion, these in this, those in that direction . . ., in the fourth, the sons are commanded to create their likenesses. They will suffer and cause suffering. This is the first fight . . ., the older wheels turn downward and upward . . ., the mother spawn fulfills the cause. There were battles between the creators and the destroyers and battles for space. The seed appeared and continuously appeared anew."[30]

It is understandable that in 1888 the scholars were unable to extract anything intelligible from the text. Today we know better—the zeitgeist has changed. Personally, I had the opportunity on several occasions to marvel at large libraries of Sanskrit texts in India, of which just over 10 percent had been translated into English. One of these libraries is located on the outskirts of Madras, India, and was founded more than one hundred years ago by Dr. U.V. Swaminatha (1855–1945), a Sanskrit scholar.

Today, the library contains 28,000 volumes of Sanskrit texts. Only 3,200 of these have been translated into

English. The scriptures—much is known from a rough overview—contain a significant part of early Indian history, including parts of the Vedic texts. The great volume of these texts is almost unimaginable for us Westerners. The Rigveda alone contains 1028 hymns, the national epic Mahabharata 160,000 verses, the Ramayana 24,000 so-called "Schloken," and the Puranas another 310,000 verses. Compared to this stream of information, our Bible is very short. The Indologist Professor Dileep Kumar Kanjilal (born in 1933) translated many of these ancient writings and came to the unequivocal conclusion: in ancient India, there were descriptions of flying chariots, so-called *vimanas*, but also of huge space habitats and battles among gods. Some of these battles took place in outer space.[31] Such space battles are also described in the book *Sauptika Parva* of the *Mahabharata* (Section XV).[32] The weapons used were able to break apart entire planets. Today we are very close to developing such technology.

Or is Venus, as Imanuel Velikovsky suggested, the actual perpetrator of the disaster in our solar system? Did the later planet break away from Jupiter and thereby create the asteroid belt?

Can the annihilations by fire described in the myths be explained in a completely natural way, and is there no necessity for planetary wars of any gods?

No, that does not work. The Mayan astronomy refutes Velikovsky—and also confirms the wars of the gods. One must know that the Maya in Central America possessed an incredible astronomical knowledge. This is clearly stated in the *Dresdner Codex* (*Codex Dresdensis*). It is one of the three manuscripts that survived destruction by the white conquerors. The *Codex* has been at the Saxon State Library in Dresden, Germany, since 1739. At that time, the librarian Johann Christian Götze brought the old Mayan manuscript from Italy to Dresden.

Eleven pages of the *Codex Dresdensis* contain astronomical calculations concerning Venus. The Maya calculated the Venus-year to 583.92 days. Two pages of the Codex deal with the orbit of Mars, four with Jupiter, including its moons. Eight other pages are dedicated to the Moon, Mercury, Jupiter, and Saturn. The Maya knew the planets of our solar system. In addition, several pages of the *Codex Dresdensis* report "about battles between the planets."[33] The Codex lists astronomical points of reference

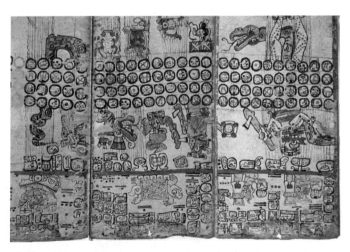

Codex Dresdensis—one of four surviving originals from all Mayan writings, from the time before the Spanish came

that occur only every six thousand years. Venus's orbital data were so well known to the Maya that they differed by only fifty-two minutes after one century. British astronomer Professor Michael Rowan-Robinson of Imperial College, London, England, said: "Such accuracies have only been achieved in Western astronomy in modern times."[34] And the American archaeologist Sylvanus Griswold Morley (1883–1948), who spent many years researching the Mayan land, discovered the city of Uaxactun, and led the excavations in Chichén-Itzá, noted: "The ancient Maya were able to determine each date of their chronology with such precision that the calculation was only successfully repeated after 374,440 years—an intellectual feat for any chronological system of ancient or modern origin."[35]

The Maya knew Venus well. It did not—as Immanuel Velikovsky suggested—emerge from Jupiter around 680 BC. And what did the Maya know about the asteroid belt? The Mayan people accomplished something completely impossible: living during one year with *two* calendars simultaneously. One of the Mayan calendars had 365 days, like our calendar. The second calendar had only 260 days. They called it Tzolkin. One week in this Tzolkin lasted 13 days. The whole year included 20 such weeks: $13 \times 20 = 260$.

They called their other calendar Haab. It contained 18 months with 20 days each. $20 \times 18 = 360$, plus 5 additional days. But really, why should a people live with two calendars at the same time? A 260-day calendar was of no use in practice. It was not applicable to the seasons; to sowing and harvesting. Why have this confusion?

The Exploded Planet

For the Maya, the 260-day calendar was considered a "divine calendar." All religious festivals were celebrated according to this calendar. It had been the gods who introduced the Tzolkin. Why was that? Presumably, they wanted to always remember the orbit of their original home planet. Had a celestial body ever orbited the Sun between Mars and Jupiter, this would have had to happen every 260 days? How do you arrive at that number?

Johannes Kepler (1571–1630) is the father of planetary laws. He was a philosopher, mathematician, and astronomer and is still, 400 years after his death, one of the most famous German scientists. He was the first to understand the relationships between planetary motions and formulate them in Kepler's Third Law, which is of great importance to astronomy. Kepler understood that the planets had to spin around the Sun in an elliptical orbit. He calculated their orbits, including Earth's orbit, with incredible precision. His formulas take into account the size of the respective planets and the gravitational pull of the Sun. According to Kepler's laws, a planet would have had to orbit between Mars and Jupiter—but there is only the asteroid belt, the remains of the former planet of the gods.

An exploding planet causes a great devastation throughout the solar system, and not just for a few days. Centuries passed before astral bodies had assumed their new celestial orbits. Whole swarms of "glowing stars" broke through the earthly atmosphere. Stretches of land soon burned, lakes boiled, and rocks became glass. The desperate people sought shelter in grottos and caves. Then our ancestors began a superhuman task. They built huge shelters, thousands upon thousands, all over the world. Some chiseled miles of subterranean tunnels through the rocks and others managed to shelter hundreds of thousands of people in subterranean cities! How can that be proven? One at a time.

Euboea is the second largest island in Greece and is 175 kilometers long and 45 kilometers wide. It is separated from the mainland by the long gulf of Euboea. At a height of 40 meters, a bridge crosses the narrowest strait on Earth in Chalcis.

In the south of Euboea there are twenty-five so-called "dragon houses." Below them are huge dolmen-like buildings with megalithic walls and ceilings. The word *mega* is of Greek origin and stands for "big, powerful." *Lithos* is "the stone." Megaliths are thus big stones. Since time immemorial, the buildings have been called *dragon houses* because they are associated with flying, fire-breathing beings of the firmament. In ancient times, the buildings are thought to have protected people from the dragons. In the 19th century, researchers suspected a variety of uses. They are seen as precursors of Greek temples.[36, 37] Astronomical meanings were also assigned to the dragon houses. The Greek astronomer Theodossiou even suspects a connection with the star constellation Sirius.[38]

There is speculation about the age of the buildings, but one thing is certain: they have been there since the Stone Age, because they were built of stone and nothing but stone. Greek researcher Vasilis Kalalougas inspected all twenty-five Euboean dragon houses and presented them in an impressive illustrated book with black and

The whole family of Echnaton was longheaded—even the small babies.

Longheaded Gods and priests from ancient Egypt on different steles in the museums of Vienna and Berlin

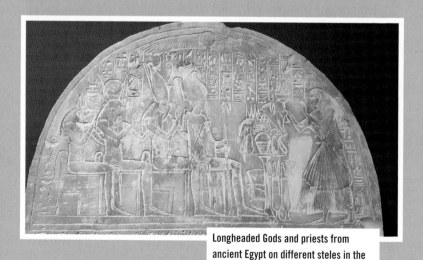

Longheaded Gods and priests from ancient Egypt on different steles in the museums of Vienna and Berlin

An impressive collection in the Museo Juan Navarro Hiero in Paracas

One of the best specimens from the Museo Juan Navarro Hiero in Paracas

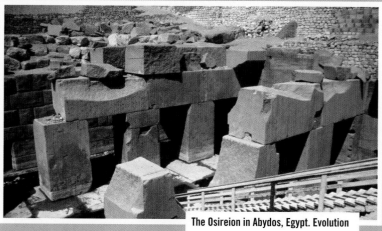

The Osireion in Abydos, Egypt. Evolution of technology is upside down. The oldest blocks are the largest, the later stones are just small.

On top of Mount Ochi exists the most impressive Dragon house—gigantic stone blocks, perfectly cut and with poly-surface edges. Ramon Zürcher pictured.

On top of Mount Ochi exists the most impressive Dragon house—gigantic stone blocks, perfectly cut and with poly-surface edges.

Gigantic halls and tunnel systems in China

Gigantic halls and tunnel systems in China

More from the gigantic halls and tunnel systems in China

More from the gigantic halls and tunnel systems in China

Underground cities in Cappadocia, Turkey

Chincha temple ruins

white pictures.[39] In our time, Giorgio Tsoukalos, the presenter of the television series *Ancient Aliens*, together with my never-tiring secretary Ramon Zürcher, explored these

Drago Spiti—The mysterious dragon houses of Euboea, Greece

enigmatic buildings. (I am grateful to him for the excellent pictures in this chapter.)

"The common feature of these houses are the gates, where the frame usually consists of two huge stone plates," notes Vasilis Kalalougas. Three of the dragon houses of Palli-Lakki are located on the western slope of the 682-meter-high mountain Kliosi. Although they are classified as smaller dragon houses, their construction is typically megalithic. The north and south houses measure 12.4 × 6.2 meters. In between there is a 7-meter-wide courtyard and, on the east side, the third house. Here, the walls, together with the ceiling, consist of rounded, heavy stone slabs, one on top of the other, as the structure becomes narrower toward the top. Who, during this ancient time, came up with the idea of building dragon houses on mountain slopes—and what was the point? In addition, several Stone Age families must have had the same thought—after all, there are twenty-five dragon houses on Euboea alone.

The most imposing of all these structures stands on Mount Ochi. Located on the slope of the same mountain, only 500 meters away, is the monastery of the prophet Elijah. He went to heaven in the Old Testament in a fiery chariot. (See 2 Kings, Chapter 2, Verses 11–15.) It happened when Elijah spoke to his favorite student Elisha:

> *Then suddenly appeared a fiery car with fiery horses from the sky and separated the two. So Elias went to heaven in a whirlwind. As Elisha looked on, he cried out, "My father, my father!" Then he did not see him anymore. Elijah's coat fell down to the ground, Elisha picked it up and turned back to the banks of the Jordan . . .*

On top of Mount Ochi exists the most impressive Dragon house—gigantic stone blocks, perfectly cut and with poly-surface edges.

The ascension of Elijah most likely did not take place on Mount Ochi on Euboea, but you may ask why a monastery of Elijah with a total of four cells was built right there on the mountain of dragon houses.

In addition, the word *Ochi* is a variation of the ancient Greek *Ohevo* and once meant "riding, driving." Does this mean to ride or drive to heaven?

The Ochi dragon house sits at an altitude of 1,386 meters between two mountain peaks (geographical position: 38 ° 03 '31" north + 24 ° 28' 03" east). It consists of mighty ashlars, cyclopean stone beams, and ceiling plates.

Its measurements are 12.7 × 7.7 meters. The blocks above the entrance were cut at right angles. It was quality craftsmanship. Neither the execution nor the tools used would lead one to believe that these clean-cut blocks stem from technically uneducated humans of the Stone Age. Here an irrepressible will to protect oneself from danger must have prevailed. Even today, as you stand –12700413385000 in a dragon house, a sense of absolute security comes over you. Some of the stone blocks and plates have a thickness of 50 centimeters. The overarching ceiling monolith is 4 meters long, 2 meters wide, and 34 centimeters thick. Here you feel safe and protected from all storms, lightning, and possible dangers from space. Was this the real reason and the driving force that made the builders carry out this titanic work?

Were these built for protection from something powerful that was not from this Earth? The Stone Age humans knew that flying dragons did not exist, and with their weapons they were able to defend themselves against many animals, especially in the mountains. But only shelters helped against the dangers of force majeure. These were places where the families could wait until the sky had calmed down again.

Such shelters do not only exist on Euboea, or even only in Greece, or even only in Europe, but they number in the tens of thousands on our large, vast Earth. Independent from each other, our ancestors must have been overcome by the same idea: "Build passage graves and great dolmens!" And all miraculous megalithic constructions were, without exception, built during the Stone Age. Skeletons have been found only in the few passage graves and great dolmens. The situation is different with the smaller dolmens. Most likely, they were supposed to be simple tombs. At some point, our Stone Age ancestors could no longer look at the body of a loved one being devoured by vultures and hyenas or mauled by wild boars and other creatures, so the clan began to bury their dead. I could imagine that until then, those who remained behind stared with gleaming, sad eyes at the spot on the ground where the admired person once lay. Was that all, was there really nothing left of him? One carefully held the few scraps of fur, tools, or works of art left behind by the deceased. Gradually, a veneration or cult of the dead arose. Could the person who had gone through life live on somewhere else, or maybe even return? Did the caterpillar not pupate to wake up as a butterfly in the spring? Would the person returning from the realm of the dead ask for his weapons, tools, clothes, and favorite objects? So they began to solemnly bury those who had passed away.

The ground was hard, the stone tools were inadequate, they could not make a deep crypt, and so animals still brought body parts back to the surface. The idea of laying stone slabs over the burial sites was developed, and finally, based on this idea, they began to build the dolmens.

Nothing against these seemingly reasonable considerations, but when it comes to large passage graves or giant dolmens, they fall short. Why? The general opinion is that gigantic dolmens such as those of Newgrange in Ireland (51 kilometers northwest of Dublin or 15 kilometers west of the town of Drogheda in County Meath) were erected as funerary monuments for powerful princes. But when a child was born, no one could foresee whether he or she would ever become a hero or a kind of superman or superwoman. The grave, however, had to be planned and built. Extensive calculations and measurements had to be made in advance and the terrain had to be leveled. Afterward, the megaliths had to be taken out of a quarry, processed, transported to the building site and assembled perfectly,

The famous Newgrange in Ireland—it was never a grave mound.

which required at least two generations of people spending time on the task. So the grandfather would have to commission the funeral crypt for the future grandson. The builder could not even know if the hypothetical heir would even become a hero or heroine or perhaps die in a distant land. Even if he had the facility built for himself, where are the graves? They do not exist anywhere. People who live more than an average life usually have their names immortalized. But no names have ever appeared in any of the great dolmens.

Today, rich people build nuclear bunkers under their properties or in domes of rocks. The underlying idea is always safety. Something terrible could happen against which we have to protect ourselves. As in the case of the dragon houses on Euboea, I suspect that these great dolmen structures were also safety structures. In case of a fall, one could have sought shelter there.

Newgrange, Ireland

The Gavrinis tumulus in French Britany

The term *dolmen* was introduced by Frenchman Théophile M.C. de la Tour d'Auvergne (1743–1800). Originally, it meant a grave made of lateral support stones and one or more ceiling plates. But the great dolmens such as Newgrange in Ireland, Gavrinis in French Brittany, and the Cueva de Menga (Menga Dynasty) with a 180-foot ceiling stone, or the Tholos of El Romeral not far from it, both located in Andalusia, Spain, (accessible via the N 342 from Granada to Archidona), do not fit the original meaning of the word *dolmen* at all.

A side note: in the summer of 2018, the No. 8 highway between Bern and Interlaken in Switzerland was widened. The construction was disturbed by a so-called "foundling," which is a stone that remained lying in the area millions of years ago when the glaciers retreated. The road builders had to move the obstacle a few meters. For the claws of the construction tools to grab a hold of it, several holes were drilled into the granite block. A crane device with two massive crossbeams lifted the over 300-ton block and steered it 3 meters to the side to make way for the new highway. In view of our current technical

Cueva de Menga in Spain, built of megalithic stone blocks

effort, the question is: how did Stone Age humans transport blocks weighing hundreds of tons, some in hilly terrain, such as in Baalbek, Lebanon? There, blocks with a weight of up to 1,600 tons were moved.

Of the smaller dolmens that originally served as graves, there are hundreds of thousands worldwide. There are at least 5,000 of them in Germany alone: 190 are located in Schleswig-Holstein, 280 in the Hanover area, and 63 in the Oldenburg area. In Germany they are also called *Hünengräber* ("tombs of giants"), even though no bones were ever found in any of them. East of Bremen, near Wildeshausen, stands a large dolmen, which consists of a total of 134 megaliths and is a whopping 104 meters long. Who built the site, for whom, and when it was built remains a mystery, because there are no names or objects that could be assigned to specific persons. Things are a bit different in Russia and especially in the Caucasus. In western Caucasus alone, there are 3000 dolmens, all built with precision and

A massive technical project to move a heritage erratic boulder for a highway expansion

partially assembled at right angles, as if the builders had worked with files and rotary cutters. Surprisingly, there are no quarries within a radius of 100 kilometers. Bones have been found in several Caucasian dolmens.

But as demonstrated in the museum of Gelendschick—located on the northeastern Black Sea coast in the foothills of the Caucasus—the bones and megalithic structures date from completely different times. The dolmens had already existed for a long time before they were much later used as graves. In the Caucasus, a local legend has it that dwarves and giants lived peacefully side by side long, long ago. The dwarfs were so small that they used rabbits as mounts. For their part, the giants would have set up the dolmens as dwarf dwellings. But they didn't do this for free. The dwarves regularly provided the giants with plants that altered their minds. A few centuries later, the giants had become imbeciles. They trampled the dwarfs to death and eventually became extinct themselves.

Denmark is dotted with many dolmens. There are 119 of them on the island of Møn alone. Among them are some great dolmens, such as Skaglevaddyssen (26 meters long) or Cronsalen (102 meters long). The age of primeval buildings is unclear. They could have been made anywhere from 10000 to 6000 BC; anything is possible. But even here, the few bones that were discovered are much younger.

The same applies to Sweden. Dolmens exist in the provinces of Halland and Småland, on the Baltic island of Öland, in the county of Kronoberg (near Ljungby) and in Västergötland in southwestern Sweden. Scientists don't know for certain who the builders were and what the original reason for the megalithic structures was.

This statement applies, however, worldwide. Who knows of dolmens in North Africa? They exist in Tunisia, Algeria, and Morocco. They exist in Lixus, which is

A dolmen in Morocco at Lixus

located 100 kilometers southwest of today's city of Tangier. And far from Europe we see the same story: there are dolmens in Japan and Korea.

The sites of Gochang, Hwasun, and Ganghwa have even been included in UNESCO's World Heritage list for Asia and Oceania. There are 35,000 dolmens in Korea alone. The same goes for India. East of the city of Dharwar in the Indian state Karnataka lies a plateau with the hill Durgadad. And there, presented like a panoramic view of ancient times, lies an incredible world with hundreds of dolmens of different sizes. Five-meter granite slabs rest on three-meter-high monoliths as if they were tables for giants. In between, overturned menhirs, oversized stone slabs, and remnants of stone circles can be admired. It appears as a chaotic open-air museum.

It's unbelievable: a planet full of dolmens and no science to explain why. Although the prehistoric Europeans built dolmens by the thousands and inspired each other in their construction, the question remains, why did the distant Indians, Koreans, North Africans, and Colombians do the same? And if even smaller sites were later used as graves, why not build something much simpler? The weight of many of the individual blocks of those dolmens in the Caucasus is 15 to 30 tons. Grave sites could be secured with much smaller, lighter stones. And these gigantic graves could not have been dedicated to once admired and revered people because, as I emphasized, no names were engraved on any of them.

The Frenchmen of prehistoric times were the best of these examples with their great dolmens. Brittany has rows of Carnac stones (menhirs) that can be seen for as far as a mile, which, as we now know, all reflect geometric messages.[40] It is also the place where we find the sites with the largest dolmens on the planet. One of these is located in the village of Essé, between Vitré and Châteaubriant in Brittany.

The large-scale structure carries the name *La Roche-aux-Fées* (meaning *rock of the fairies*) only because it was impossible for many to imagine how Stone Age humans would have transported such weights. The construction is 21 meters long, 6 meters wide, and 4 meters high. It consists of twenty-six side stones, so-called *orthostats*, and eight ceiling stones, each weighing up to 45 tons. Like many great dolmen sites, La Roche-aux-Fées is astronomically oriented. On the day of the winter solstice, on the 21st of December, the sun rises exactly above the center of the dolmen entrance. Is this a former tomb? With ceiling stones weighing up to 45 tons, the word *tomb* becomes silly. As elsewhere, neither skeletons nor engraved names were discovered here. It is becoming more and more obvious that the great dolmens found worldwide served as shelters.

The next of these amazing creations from the Stone Age is called *Table des Marchand*, meaning *table of the (merchant family) Marchand*. It is located in the village of Locmariaquer in Brittany, France, and is covered by an 8-meter-long, 4-meter-wide plate that weighs 50 tons. It remains an absolute mystery in our enlightened computer age how people of the Stone Age could lift a 50-ton plate over the side stones. And why did they ever undertake such a superhuman task? For a grave, lighter plates would have served the purpose. But the Table des Marchand has an additional curiosity. In 1979, archaeologist Dr. Charles-Tanguy Le Roux discovered strange engravings on one of the huge ceiling stones that seemed somehow unfinished or half-finished and then destroyed. They showed neither names nor writings, but a large, ax-like object and pointedly approaching lines. The same Dr. Le Roux also investigated another great dolmen called Gavrinis (also in Brittany).

He immediately noticed a similarity between the engravings of Gavrinis and the Table des Marchand. In fact—a comparison of the two dolmens confirmed the assumption: the cutting points of the engravings fit together perfectly. The ceiling slab of the Gavrinis dolmen, like that of the Table des Marchand, had originally been a single, gigantic block! It seems obvious that the engravings must have been done in the quarry. Otherwise, one half of the engraved block wouldn't have ended up in Gavrinis and the other half at the Table des Marchand in Locmariaquer. It seems to me that the builders of both dolmens were suddenly unconcerned with the engraved message. The aim was to have huge, solid ceiling stones available for both dolmens as quickly as possible. What immediate danger did they think they were in?

Only 2 kilometers from the already mentioned Cueva de Menga lies the site called Tholos de El Romeral (also called Cueva del Romeral). This Spanish location measures 44 meters in length and contains two transverse spaces each 10 meters long. Again, no bones were found, but the bottom of both chambers was covered with a thick layer of compact black ash. It cannot have originated from any kind of fire that once flared here, because the ceiling of the rooms shows no traces of smoke. It is possible that ash once also covered the soils in other great dolmens far away. But the new users of the old sites could have cleaned up the dirt.

Why did people of the Stone Age create underground temples? They exist worldwide; for example, in Central America, like the ones created by the Maya. Such underground sites were found under Palenque and under Chichén-Itzá. The same is true for Xochicalco, which lies at the foothills of the volcano Ajusco near Mexico City. The main temple is the step pyramid of La Malinche, and it is decorated with symbols of a flying snake—or could it be a dragon? An astronomically oriented sanctuary can be found underground. Under the gigantic pyramid city Teotihuacan, which is located about 50 kilometers northeast of Mexico City, as well as in the distant Tiahuanaco in Bolivia, subterranean sacral systems have been found. The same applies to the impressive temple complex of Chavin de Huantar in Peru, and, of course, to other sites in Europe. The best-known subterranean temple is the Hypogeum on the island of Malta. With all these underground systems, the question is why religious communities have to create sacred places underground in addition to their imposing structures on the surface? I suppose this was not so much about secret rituals, which were celebrated in closed rooms on the surface—the main driving force was security. Were they wondering what they would do if they could not perform their ceremonies because the sanctuaries on the surface had been destroyed?

The step pyramid in La Malinche with the feathered serpent

Also, you find a subterranean astrological sanctuary.

In addition to the man-made dolmens, there are also the so-called *paradolmens* (or *pseudodolmens*). These are mixed systems of natural, local megaliths and artificially added rock slabs. The best known sites can be seen in Catalonia, Spain; in Luguria and Tuscany, Italy; both Italy and in France. With paradolmens, it can be difficult to argue that they were originally planned as graves. Why? The graves of princes or ethnic people were not allowed to be built just anywhere; that is to say, they were not built far away from their own tribal territory. There might have been foreign customs, and the location of a grave site might have been out of the control of the tribe, but people were generally buried close to where they lived. Paradolmens, however, often emerged far from people's settlement areas.

Forty-five years ago, I reported on a cave system in Ecuador, South America.[41] And because I was so vehemently attacked, I explained the story again thirty-five years later.[42] Yes, that tunnel system in far-off Ecuador actually exists. Here I show a picture of the entrance and one of the insides of the cave complex. The entrance to the subterranean world is still guarded today by the indigenous tribe of the Shuar. The white Ecuadorians called the native people Jivaros, and before 1950, they were known only as headhunters. And where exactly is the main entrance? For future researchers who can afford the protection of the Ecuadorian army, here is the geographical position: 77 ° 47' 34" West and 1 ° 56' 00" South.

The question that I was never able to answer, namely why the Ecuadorian tunnel system was built, also applies to the paradolmens and systems. The Ecuadorian tunnel system is a mixed system of natural caves and man-made crossroads, and so is Chinkanas in Peru.

The word *Chinkana* refers to a labyrinth of artfully carved rock tunnels and of natural caves that intersect, turn, bend, and cross above and below each other in all

Side tunnels of the Tayos caves in Ecuador

directions. One of the entrances to this labyrinth lies just below the main altar of the Santo Domingo church in Cuzco, Peru. Only a few experts dare enter the vaults with breathing masks. Even the annalist Garcilaso de la Vega (1501–1536), son of conqueror Capitan Sebastián Garcilaso de la Vega, ventured only as far into the vaults as the daylight reached and the candle stayed lit.

These subterranean systems, connecting tunnels, halls, ceremonial rooms, and so forth existed during ancient times. But we don't know *who* created these tunnels and spaces, and by *what means* and *when* they were created. But we might be able to answer the question as to *why* they were created.

As a Swiss citizen who lives in the Swiss mountains, I know some of the kilometer-long military installations from World War II. At they time they were built, my compatriots feared a raid from Nazi Germany. Bunkers, hospitals, ammunition depots, crew quarters, sleeping rooms, and also halls for tanks and war material were all underground, blasted into the mountain rocks. The entrance to one of these facilities is less than 2 kilometers from my house and is now open to the public. I have visited it, and you step back and learn to be amazed again. All rooms are interconnected by tunnels. In the area of the Grimsel water dam in the Bernese Oberland alone, there are over 50 kilometers of tunnels one can drive through. How will archeologists of the future judge these rock labyrinths in 1,000 years? Will they classify them as tombs, crypts for ritual ceremonies, or as a safe site for the persecuted? No one will think of the most logical explanation for them: safety.

Thousands of years ago, there were no world wars and no one had to get to safety from the threat of nuclear bombs or fighter planes. The enemies from neighboring countries were all familiar. They would also have known about the underground facilities, because such achievements as the creation of labyrinths underground don't go unnoticed. Enemies would only have to block the entrances to the tunnels and starve their opponents. Does this argument not also apply to the military installations of Switzerland during the Second World War? No. All entrances were in the range of gun positions, which were placed in the surrounding mountains. Troops trying to block the openings would have been shot from a distance. Millennia ago, however, there were no long-range guns that could protect the tunnel entrances from besiegers. Moreover, in the military tunnels there is not just one entry or exit. There are plenty of them.

But subterranean facilities from prehistoric times do not exist only in distant lands. They are present at one's own doorstep. In Styria, Austria, alone, around 800 so-called *Erdställe* (literally: "earth stables") have been investigated. In his book *Geheimnisvolles Österreich* (*Mysterious Austria*), my colleague Reinhard Habeck, one of the researchers whose statements can be relied on, explains the term: "Even the word 'stall' (literally 'stable') seems strange. The word here stands for 'spot,' 'place' or 'location' and has nothing to do with a building for housing animals. Upper Austria is perforated like Swiss cheese with such artificially created 'earth stables' . . . But these places can be found far beyond this area. Lower Austria,

Bohemia and Slovakia as well as Poland and Hungary, but also Bavaria, France and Spain, point to this mysterious underworld. There are more than 2,000 of these sites in Central Europe."[43]

These estimated 2,000 installations are those in Greece, Turkey, and Israel. Even decades ago, Turkish and Greek researchers showed me such subterranean "dwarf tunnels." At that time, I did not pay much attention, because the general view was that these were the passageways of animals. Now, that opinion has changed. The Austrian researcher of prehistoric times, Dr. Heinrich Kusch from the Karl-Franzens-Universität in Graz, together with his wife Ingrid, has been researching several of these curious Erdställe for decades. There are supposed to be hundreds of them, and quite a few of them can be reached by walking through tunnels. The results of the age determinations, which were created by Dr. Kusch, are absolutely astonishing. A ceiling plate in the Kandelhofer Erdstall near Puchegg in Styria, Austria, showed an age of 23,965 years (+/− 694 years). Processed stones in several other places were dated at an age of 10,000 to 12,000 years.

The dating of the objects was carried out with the so-called terrestrial cosmogenic nuclides (TCN) method. TCN is a chemical-physical measurement method that has been tried and tested in archeology and geology. Of course, the approximate age of 24,000 years of the processed Kandelhofer stone slab has been criticized by skeptics: it could be that the slab weathered for a long time outdoors and was only much later moved by human hands into the Kandelhof Erdstall, but this is contradicted

The Exploded Planet

Perfectly shaped tunnel systems in Austria

by Dr. Kusch's thorough research. He suggests that the stone slab was fitted precisely, down to the centimeter, into the space delimited by rock walls, and thus only specially manufactured for this place. In the so-called *French cave*, his team was able to pinpoint the date of creation for a processed ceiling plate to 10,893 BC (+/− 393 years), and for another plate in the Grubergang to 10,382 BC (+/− 288 years). In his books, which he published with his wife—*Tore zur Unterwelt* (*Gates to the Underworld*) (2009) and *Versiegelte Unterwelt* (*Sealed Underworld*) (2014)— Dr. Kusch states that these stones are not boulders that a glacier left behind millennia ago. Furthermore, surfaces of local boulders from earlier ice ages were dated to around 55,000 years apart from the processed stone slabs. This age corresponds to the true age of boulders in this region because in the last ice age there were no glaciers in this area that could reduce the age of the stones. Also, weathering of the rock would trigger an older age.

However, the classical opinion of some researchers dates the Erdställe to a time period from about 800 to 1200 AD, although this has not been proven by any scientific evidence. The archaeological finds and organic materials used to arrive at this date were put inside the earth stables at a later date. There are many hundreds of such finds, which originate from the Bronze Age or from the Neolithic age, but which have also been put there by humans. This means that the earth stables with the prehistoric finds were already present at that time.[44]

So the classical explanation is incorrect. Also, the time period from 800 to 1200 AD is, more or less, well known.

The Exploded Planet

For example, on November 27, 1095, Pope Urban II issued his call for the Crusades. In July 1099, Jerusalem was conquered, and in 1207 Pope Innocent III ordered the annihilation of the Cathars. Yet, the construction of earth stables is not noted in any local chronicle. But tens of thousands of workers across Europe would have to have been involved. The vastly different dates make everything even more confusing. Additionally, several of the earth stables are located under monasteries, but the monastic chronicles contain nothing about them. Why were the Erdställe created?

Reinhard Habeck comes up with every conceivable approach to answering this question. Did they serve as refuge? If so, for whom? There was not much room in the earth stables. Were they storage areas? Why did we not find any decayed stocks? In addition, the entrances are narrow. Did people dig for raw materials? Habeck rejects the assumption and refers to the fact that for a "mountainous exploration," *one* tunnel shaft would have sufficed. "But at a distance of only a few meters, elaborate and irrational tunnel systems were constructed."[45] Did children create the earth stables as playgrounds? Forget it! The earth stables exist in different rock layers.

Did they serve as rooms for cults? Were they roaming places for the spirits of the dead? Did they serve as a place for healing magic where patients would become healthy again? Were they hiding places for valuable items? I don't think so; how could this idea have suddenly occurred to farmers all over Europe? There are now great organizations, clubs, research groups, books, and newsletters dealing with the mystery of earth stables. All possible reasons for the construction of these underground facilities have been thought through, and none is convincing.

In northern Styria alone, says Kusch, "there are more than eight kilometers of walk-in corridors at different sites. And the absolutely inexplicable thing about it: At least in some tunnels, machines have been used for drilling."Kusch assures us that his team scanned these corridors' spaces to an accuracy of 0.2 millimeters and found that "there are only deviations of 16 millimeters from the tool tracks on meter-long routes."[46]

What may have happened unknown millennia ago in the territory of Styria, Austria? What kind of technology was used? Who operated the machines and where are their remains? And every month, more of these inexplicable corridors, shafts, and underground spaces appear. From the year 2012 on, representatives of the Church connected with Dr. Kusch and provided him with information about the likely purpose of these old underground facilities.

For example, Frenchman Luc Stevens reported the discovery of a 120-meter-long earth stable in Prinçay (a municipality of Availles-en-Châtel-lerault, Vienne district, France), and during construction work in the parish garden of the municipality of Aying in the district of Munich, an earth stable with a 40-meter-long main tunnel and a spacious back area was discovered.

In his book *Überirdische Rätsel* (*Celestial Riddles*), Reinhard Habeck reports an additional mystery that was detected in one of the earth stables.[47] Under

Klosterneuburg, a town northeast of Vienna, a labyrinth was discovered "with cellars, corridors, and vaults." A wooden chest with black stones was also found. "When sunlight strikes, the stones turn a bluish tint, and when they are illuminated in the dark with ultraviolet light, they reveal glowing symbols on their surface like the omega sign." Habeck wonders if the black stones are the product of "magical practices." Perhaps they are, but why do their chronicles not show any report of building a labyrinth under their own monastery walls? If all this had arisen in the twelfth century at the same time the monastery was built, the monks would probably have written about it and about the purpose of the stones, which reveal their message only under ultraviolet light. And the magicians who mysteriously engraved invisible messages on the stone surfaces during the day were supposed to know that long after their time, someone would come up with the absurd idea of illuminating the stones with ultraviolet light!?

The world under the earth is more spacious than we think. Luc Bürgin, editor-in-chief and publisher of the magazine *Mysteries*, tells us in his *Lexikon der verbotenen Geschichte* (*Lexicon of Forbidden History*) about a case from the Nazi era.[48] Toward the end of World War II, a group of soldiers hid some valuable objects under the town of Bühl (near Baden-Baden, Germany). The objects were not in tunnels that they dug or blasted, but in tunnels that already existed. To this day, it is unknown what the soldiers hid, because the entrance to the underground world was walled up by the authorities.

Miles of corridors and spacious underground halls also exist far away from Europe. For example, the astute journalist Luc Bürgin visited a gigantic cave labyrinth in distant China several times. In the southeast of the gigantic country, in Anhui Province, not far from Huangshan, lies the largest man-made bunker system in China. In his book *Chinas mysteriöses Höhlenlabyrinth* (*China's Mysterious Cave Labyrinth*), Bürgin shows over a hundred crisp images from thirty-six caves—all carved in ancient times "from quartzitic-bound, partly coarse-grained, partly medium-grained sandstone."[49] Bürgin describes "multi-level caverns" with the type of massive support columns, "that three adult men can hardly encircle with their arms." The gigantic sites are now mostly under water. The largest hall yet uncovered resembles "a gigantic underground palace or huge cathedral." Bürgin and his companions Floyd and Jenny Varesi walked through miles of corridors and rooms reminiscent of mausoleums. Only a small part of the endless tunnels has been made accessible to tourists. Structures with several floor levels, all carved out of the rock by human hands, are under water. And again and again the researchers came across curious engravings. These are not writings, but serpentine lines carved in the rock, crescent-like notches, or tapering grooves. Bürgin knows the subterranean Hypogea site on Malta and calls it a "speck" compared to China's underworld.

More underground sites have been discovered west of the Chinese province Zhejiang. Zhejiang is located in southeastern China on the coast of the East China Sea.

The Exploded Planet

Over 70 percent of this province consists of mountainous terrain. Forty-five man-made domes of rock have been discovered; five of them "with pillars of up to twelve meters in height."[50] The Zhejiang site is only about 200 kilometers from the site in Huangshan. It cannot be ruled out that both systems are connected. In the meantime, Bürgin visited the tunnel and rock world of Zhejiang with a group from the Research Society for Archeology, Astronautics and SETI (A.A.S.) (see last book pages). The researchers marveled at underground halls of up to 3,000 square meters in size and 27 meters in height.[51] This is equivalent to a six-story building.

In one of the grottos, an engraved script read "Emerald Cave." And this very term appears in a poem from the Song era. (The northern Song period lasted from 960 to 1279 AD; the southern one from 1126 to 1279 AD.) Even

Gigantic halls and tunnel systems in China

then, the poet writes, the cave was very old. So the subterranean dome already existed before the Song era.

I have known Luc Bürgin since he was a boy. He became a journalist and the eloquent editor-in-chief of a newspaper in Basel. Bürgin is one of those men who doubts everything and thinks nothing is impossible. The young man became a globetrotter and lateral thinker, writing only about things he can prove. That's why I believe him when he talks about 800,000 cubic meters of rock mass that had to be carved out of the rock for certain caves, or about the mysterious fact that seven of the caves from a "bird's-eye view resemble the constellation of the Great Dipper."[52]

They are similar to the great dolmens, the temples under the earth, the labyrinths of underground passageways in Ecuador or Peru, or the smaller earth stables in Europe.

Nowhere have the builders shared their working methods, nowhere do we find inscriptions with a date, and nowhere do we learn the purpose of their laborious work. We do not know how the people of the Stone Age hoisted 300-ton blocks over side stones, or when the worldwide drudgery got started. The Huangshan sites are located in the geographical area where, millennia ago, the legendary yellow emperor used to travel from the summit of the surrounding mountains to the stars. Did this heavenly emperor command the construction of the subterranean world? Why should people carve huge rooms and connecting tunnels out of the rock? Did the earthlings want to be close to their supernatural teachers, and was this somehow only possible underground, or was there at least a temporary danger on the earth's surface? Were there things that could only be stored underground, and if so, what were they? No traces have been found. What is certain is that China has been the land of mystical dragons for millennia. All Chinese rulers considered themselves *Tianzi*, which means *sons of heaven*. They all lived by the rules of *Tianming*, the *mandate of heaven*. And the largest temple in atheist Beijing still bears the name Temple of the Heavenly Emperors today.

We even find subterranean cities on our unhurried planet—yes, cities!—big enough for hundreds of thousands of people. Following is the story.

The word *Derinkuyu* is Turkish and means *deep shaft*. The village of the same name is located in Cappadocia, 29 kilometers south of the provincial capital Nevşehir. In autumn 1963, a farmer stored two sacks of potatoes in his cellar. When he wanted to get these potatoes weeks later, the bags were half empty. The farmer looked for rats and other animals and discovered a hole in the ground. Obviously the potatoes had fallen down the hole. Using a flashlight, he discovered a shaft in the ground, but he didn't see any potatoes when he shone his light into the darkness. Neighbors brought tools and helped him dig into the ground around the hole. The hole led to a shaft, which was clearly man-made. The following days and weeks became more and more exciting. Additional tunnels led to larger and smaller rooms, then to other floors underground. Most amazing was the ventilation. Wherever the search party crawled, there was a slight breeze in every hallway, and the temperature was the same everywhere. Hunched over, the men hobbled through a tunnel that seemed as if it would never end. After half an hour of walking straight ahead, the flashlight batteries died. Four days later, another group set out, equipped with new batteries and two lamps with dynamos. In addition, torches and candles were available for emergencies. Hours passed, and they walked many kilometers under the earth. Men and batteries were replaced, and food and drinking water was brought in. Finally, after walking many kilometers in a constant temperature, the searchers encountered a bend and a larger space. Then, they came to a flight of stairs and finally to a mighty, wheel-shaped monolith. It lay in a side niche and clearly served to seal off the subterranean world from the inside.

Back on the surface, the men were stunned to find that they were in Kaymakli, twelve kilometers from Derinkuyu. Derinkuyu and Kaymakli were just the beginning of an adventure leading to the discovery of an incredible labyrinth under the Earth. Fifty subterranean cities have now been verified, all interconnected by tunnels. UNESCO has long declared the area a World Heritage Site. Derinkuyu alone provided shelter for twenty thousand people. All cities together can house 1.5 million persons. Unknown millennia ago, when the cities were hewn out of the rock, they could not have been fast, improvised refuges. There were common rooms, sleeping areas, and even stables. All subterranean cities contained multiple floors. There are thirteen of them in Derinkuyu alone. The various living silos are all interconnected, the entrances to the outside could be sealed by large, round stone doors. At a depth of 85 meters, shafts stretched

The Exploded Planet

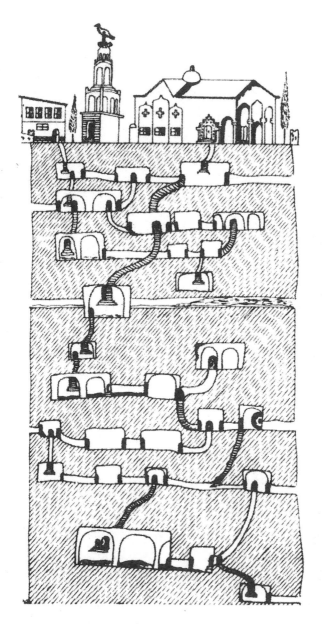

Underground cities in Cappadocia, Turkey

down to the groundwater level. A total of 15,000 of these water shafts exist in the site that have been examined so far, which also ensured even ventilation using the cool temperature of the water. Independently from the water shafts, the searchers also found pure fresh air shafts; so far, fifty-two of them. A gentle breeze still circulates today into the very last corner.

Then archeologists suspected that the underground cities were created by the Hittites in the second millennium BC. That doesn't get us anywhere. The Hittites built megalithic structures; besides, the same reasons exist as for the persecuted Christians. Any enemy could have starved the people hidden underground. The excavation of these subterranean worlds would have brought mountains of debris and rock to the surface. Once you have crawled through two of these sites, you know that these were not temporary caves. People planned and worked here for over a decade.

Recent explorations have led to new caves. In the meantime, Cappadocia alone is estimated to have up to one hundred subterranean cities. But in Derinkuyu alone, hundreds of thousands of square meters were already drilled and carved out of the rock. *The question is, by whom, when, and why?*

We realize we have the same old questions and no new answers. If there were a few tunnels cut out of the rock in a limited geographical area, so-called paradolmens, then reasons for their creation could be found, or the local traditions would tell us something about them. But the subterranean world is a global phenomenon, which certainly makes this a global mystery. People went underground or built megalithic shelters in Asia, in South and North America, in Africa, in Europe, on Greek islands like Euboea, or in the Mediterranean like on Malta. The distant Asians knew nothing about the ancestors of the Turks, and the Ecuadorians of South America knew nothing about the Chinese Middle Kingdom. (I've also read about subterranean sites in the Arctic—but I lack the evidence for that.) Global collusion did not exist, but rather a global logic. Something forced people around the world to do the same thing. It has to be about protection. This wasn't protection against attackers; that argument is moot. Were people seeking protection against volcanic eruptions? There are so many of these underground sites that are not near volcanoes. There are volcanoes in Cappadocia, but their lava would have flooded the subterranean worlds, too. Why did the prehistoric Austrians drill kilometers of the Erdställe? Why did the prehistoric Chinese hack huge, multi-story halls from the rocks? What on earth caused the indigenous people of Cappadocia to place a hundred cities under the surface of the earth and connect them by tunnels? Why did Egyptian Pharaohs drive endless shafts under the desert sands of Saqqara? Why were rooms with sarcophagi placed deep under the Great Pyramid—wasn't the pyramid itself monumental and earthquake-proof enough? What was the reason for powerful great dolmens in France and Spain with ceiling plates of up to 300 tons? Why were there dragon houses on Euboea and tunnel systems under the South American Andes? Which motives forced Stone Age people into their underground worlds? What was going on in their heads? What or who commanded: Drill! Drill! Drill!

All reasonable reasons have been considered again and again. Did people fear attacks from space? Did they want to protect themselves from greedy gods? Did they want to shield themselves from those who encircled the Earth in flying machines and demanded treasures such as gold and gems? Such flying machines definitely exist in ancient literature. I recall the Vimanas of ancient India, King Solomon's flying car, the descriptions of the prophet Ezekiel in the Old Testament, the "pearls in the sky" in the Tibetan traditions, or the barques of the gods in Egypt, along with the corresponding representations of the winged sun discs, including the pyramid texts. And yes, much has been written about prehistoric aviation by people in the know.

And yet, the same argument that applies to any exploitative gods who may have behaved like pirates applies to human invaders. "The Celestials" did not have to wait long for the humans underground to surrender. I am not aware of any traditions that report gods blocking tunnel entrances, or about groups of people under the earth who hid themselves from the gods. A great dolmen would have been easily attacked by the flying ones anyway.

After all, people just hid under it.

I suspect our ancestors were afraid of the glowing stones from outer space. They were afraid of those chunks that threatened humans for centuries from the asteroid belt. Shouldn't Earth be riddled with craters? After all, that's the case on neighboring planets and even the smallest moons.

Our planet is dotted with craters like a body dotted with a myriad of pockmarks. There are hundreds of craters. It should be pointed out that two-thirds of the earth's surface consists of water so the impacts are not visible right away. The same applies to craters that today lie under cultivated land or ancient forests. In addition, volcanic craters must be distinguished from impact craters. Finally, the Earth's atmosphere protects the planet, at least from smaller meteorites. They simply burn up. What remains are hundreds of craters caused by larger meteorites. Have you never heard of them? Here is a small excerpt from a list on *Wikipedia*:[53]

- Amguid crater in Algeria
- Brent crater in Ontario, Canada
- Campo del Cielocraters (total of 10) in Argentina
- Carancas crater in Peru
- Chicxulub crater in Yucatan, Mexico
- Chukcha crater in Siberia, Russia
- Colonia crater in Brazil

- Flaxman crater in Australia
- Flynn Creek crater in Tennessee, USA
- GweniFada crater in Chad, Africa
- Haviland crater in Kansas, USA
- Henbury craters (total of 13) Northern Territory, Australia
- Hummeln crater in Sweden
- Jabal Waqf es Suwwan crater in Jordania
- Kaali craters (total of 8) in Estonia
- Kalkkop crater in South Africa
- Kamil crater in Egypt
- Kärdla crater in Estonia
- Lappajärvi crater in Finland
- Luizi crater in Congo, Africa
- Macha craters (total of 5) in Yakutia, Russia
- Manson crater in Iowa, USA
- Morasko craters (total of 8) in Poland
- Nördlinger Ries crater in Germany
- Odessa craters (total of 5) in Texas, USA
- Ouarkziz crater in Algeria, Africa
- Red Wing crater in North Dakota, USA
- Riachao Ring crater in Maranhão, Brazil
- Rio Cuarto craters (total of 12) in Córdoba, Argentina

- Rochechouart crater in Nouvelle-Aquitaine, France
- Roter Kamm crater in Namibia
- Shunak crater in Kazakhstan
- Sikhote-Alin craters (total of 38) in Primorje, Russia
- Siljan crater in Sweden
- Steinheimer Becken crater in Germany
- Tabun Khara Obo crater in Mongolia
- Talemzane crater in Algeria
- Vepriai crater in Lithuania
- Vredefort crater in South Africa
- Wabar craters (total of 4) in Saudi Arabia
- Whitecourt crater in Alberta, Canada
- Wolfe Creek crater in Australia
- Xiuyan crater in Liaoning, China

Wikipedia alone lists a few hundred impact craters. Then there are the unrecorded craters under water, under the dunes of glowing-hot deserts, and those overgrown with jungles. Our blue planet was as much a victim of meteorite impacts as the other celestial bodies of the solar system. Why does it not hail meteorites these days?

The planets in the solar system have settled down. The chaos from the asteroid belt has abated. Anything that did not crash into the other planets and moons, and especially into the giant planet Jupiter, now follows an established orbit, bar the exceptions already mentioned. But at the time when shelters were being dug worldwide,

The Exploded Planet

showers of glowing stones must often have fallen from the sky. The crater list proves it. Again and again geologists discovered several impact craters in concentrated areas. For example, thirteen of them in the Northern Territory of Australia, eight in Estonia, and thirty-eight in the Primorsky Krai region in Russia. The barrage came from outer space and was caused by the explosion of a planet. The remnants of it form the asteroid belt and this can be proved.

Here and there, one finds vitrified sands and stones on Earth that cannot be explained by any natural heat. In the southwestern Sahara near the Saad Plateau, one can occasionally find yellow-greenish glass, also called *Libyan desert glass*. In July 1999, the British science magazine *New Scientist* wrote that so far, a thousand tons of mysterious material had been discovered, the largest chunk of which weighed 26 kilograms. Initially, it was thought that the material came from a meteorite, but there was no sign of a meteorite, let alone a crater. Even the hot gases of a celestial body that might have touched the area without striking the earth did not solve the mystery. If that were true, the locations of the desert glass should have more or less followed the trajectory, but they did not do that. The material was in an oval form, from east to west, covering an area of 130 × 53 kilometers in the sand of the Libyan desert. The glass consists of 97 percent silicon and looks like a yellow-green gemstone. In an analysis published in the scientific journal *Nature*, Dr. Spencer writes: "It's easier to assume that the stuff has fallen from the sky."[54]

Vitrification was also discovered in Death Valley, California. There are the remains of former settlements. "The whole region between the Gila and San Juan rivers is covered with remains of cities with traces of melted material. Buildings and objects must have been exposed to temperatures high enough to liquify rocks and metal. . . . One gets the impression a gigantic fire-plough had rolled over the area." Viktor Farkas (1945–2011), a very thorough Austrian journalist, quoted a colleague of Captain Walker in the magazine *Sagenhafte Zeiten* (*Legendary Times*).[55] There he also mentioned a tower in the ruins of the ancient city of Babylon, 90 kilometers south of Baghdad, which looks as if "it had been slashed in two by a sword of flames . . . Many parts of the building have been turned into glass, some are completely melted. The ruin looks like a burnt mountain."

Even in Europe, the phenomenon of vitrified walls exists. This is so on the hill Tap o' Noth in Scottish Aberdeenshire in the hinterland of the city of Aberdeen. The top of the hill is the size of a football field and is surrounded by vitrified, fused rocks in several places. An explanation for the mystery does not exist. Just as is the case with the vitrified stones in the desert of Lop Nor in the Chinese province Xinjiang.

Other vitrifications exist in the southeast of the Gobi Desert, in the Indus Valley at Mohenjo Daro, and at Sete Cidades in the Brazilian state of Piaui. Tourists can observe such rock vitrifications above the city of Cuzco in Peru. There, on the plateau on which the famous Inca wall Sacsayhuaman stands, lie blocks of rock, sometimes

The Exploded Planet

Kenko Grande in Cuzco, Peru—the stone appears to have been melted.

turned upside down, with clearly visible rock vitrifications. This is called Kenko Grande. The German engineer Dr. Friedrich Bude, owner of numerous patents and a specialist in thermodynamic processes, who investigated the traces of melted material above Cuzco, wrote to me that temperatures of up to 15,000 degrees Celsius must have been released over a short period of time.[56]

What happened here millennia ago? Humans did not produce 15,000 degrees Celsius. Or did they? After all, there was once a Bronze and later an Iron Age that followed the Stone Age. In order to release the metals from the rock, our ancestors needed high temperatures. As far as the term *Stone Age* is concerned, it does not apply globally to a specified period of time. Depending on the culture, it ended five or even two millennia before Christ—but today, Stone Age people still live on the upper Amazon. At some point a wide-awake ancestor noticed that from the chunks that had fallen into fire, something strange dripped down and froze after cooling. The oldest alloy of tin bronze was found at the excavation site Pločnik in present-day Serbia and dated around 4600 BC. Tin, on the other hand, is a chemical element; a silvery, heavy metal, and at the same time so soft that it can be scratched with a fingernail. Bronze consists of a mixture of tin and copper. Copper is a heavy metal that occurs in nature. Historically, the scientists distinguish between Stone Age, Bronze Age, and Iron Age. The starting point of the Bronze Age is considered to be 3300 BC in the regions of the Near East. It was not until about a thousand years later that people invented the production of iron, and in ancient Europe, the iron smelt was not well known before 600 BC. Like tin, iron is a chemical element. However, unlike other elements, it is not a final product of *various* alloys. Iron is found in iron ore, but also came to Earth with meteorites. And so we end up with the temperatures necessary to extract iron from the ore.

Today, this happens in blast furnaces. There, iron is melted from the iron ore. Such a blast furnace generates a heat of about 2,000 degrees Celsius.

That would be far too little to melt entire rock massifs, as can be seen above the city of Cuzco in Peru and elsewhere.

Regardless, there was no blast furnace anywhere in that mysterious Stone Age anyway.

The Exploded Planet

And let's not forget Sodom and Gomorrah. In the first book of Moses, Chapter 19, beginning with Verse 24, the annihilation of sinful places is described as follows:

> *Then the LORD rained down fire and burning sulfur from the sky on Sodom and Gomorrah. . . . Thus He destroyed these cities and the entire plain, including all the inhabitants of the cities and everything that grew on the ground . . . Early the next morning Abraham got up and returned to the place where he had stood before the Lord. He looked down toward Sodom and Gomorrah and all the land of the plain, and he saw the smoke rising from the land like smoke from a furnace.*

This extermination had nothing to do with a volcanic eruption or the explosion of an imaginary blast furnace (". . . like smoke from a furnace . . ."), or with a desert storm, or with apocalyptic visions or dreams of an epileptic, but it clearly had to do with a purposeful destruction from outer space. A gruesome weapon had to be used. I have referred to this in earlier books and more recently my opinion has been affirmed by the theologian Dr. Mauro Biglino, who is able to translate the original texts.[57]

The people who crawled under dolmens, scraped miles of corridors out of the rocks, and even built underground cities for hundreds of thousands of their fellow men and women were not afraid of other people, volcanic eruptions, heat storms, or even non-existent blast furnaces—they were hiding from glowing rock bombs from space. It's understandable that this happened independently in cultures worldwide.

CHAPTER 4

Skeletons—
Not from This World

WITH THE FOLLOWING REVELATIONS I will garner a lot of vitriol, but I can handle it. Smart alecks will refer to the whole thing as fake, and others will try to "unmask" it as a fraud. The following by experts will be refuted with "counter-assessments" by experts, and the "reasonable people" will break out in a hysterical panic in the event of any UFO case and accuse me of credulity. The know-it-alls of all stripes, who have never been able to accept the unpopular facts outside their restricted thinking patterns, will cry out loud. The carousel of pseudo-reason will be set in motion like a poison spinner. And I don't care. I'm sure of the facts, and I am familiar with the counterarguments. In the long term, it will not be possible to obscure the view of an eerie reality.

On May 6, 2017, I received an email from a well-known and very reasonable lady from Luxembourg. She wrote that a certain Mr. Thierry Jamin from Cusco, Peru, would like to contact me. It had to do with extraterrestrial mummies found in the area of Nazca, Peru. She asked me to treat the information very confidentially. The lady also sent me a black-and-white picture and excerpts from an email in French written by Mr. Thierry Jamin. Here is the translation:

> *Could you contact him (Erich von Däniken) and tell Egyptologist him we found several alien bodies? And they are from the group of the "little grey ones." This is at a secret archaeological site near Nazca, just a few hundred meters from an "earth track." I personally saw one of these mummies today. It's incredible! The group that made the discovery is asking for $100,000 (USD), but I am confident that it will be satisfied with $80,000. Perhaps Mr von Däniken knows somebody in the world who makes it possible to secure these finds. The whole thing is really unique! It would be wonderful if Erich von Däniken would contact me . . . The tomb robber named "Mario" gave me one week . . .*

My acquaintance in Luxembourg added that it was a "very hot topic" and she thought that we were very close to finding evidence for the aliens.

Did she mean alien mummies? Was this on Earth? This must be a case for the psychiatrist! Any normal person would start laughing immediately. I am not a normal person. I work with information that ordinary people do not deal with, and I have been referenced in forty-one books so far. So I know that the Belgian Padre Gustavo

Le Paige (1903–1980), who worked as an archaeologist in Chile and whose name was given to a large museum in the Chilean city of San Pedro de Atacama, had seen extraterrestrial mummies, "You would not believe if I told you what else I found in the tombs," Padre Le Paige had told a journalist. I am as familiar with the reports of alien visitors as I am with the ancient texts. That is why the news of "alien mummies on Earth" did not compel me to break out in a hysterical laugh.

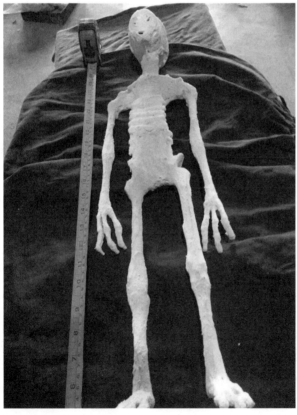

Extraterrestrial mummy or fake?

The picture that my acquaintance from Luxembourg sent did not matter. It showed a 32-centimeter figure with three fingers covered or sprayed with a white powder.

Hidden beneath it might as well be an artificial toy doll, or perhaps also a glued-together skeleton. But who was this Thierry Jamin, the man who had contacted my acquaintance in Luxembourg?

Thierry Jamin, born December 19, 1967, in Chartres, France, had searched for and found the pyramid of Pantiacolla in the Peruvian jungle in 2001. The pyramid turned out to be a natural hill, but Thierry Jamin found some ceramic pieces from the Inca era. He is the discoverer of several petroglyphs of unknown cultures. In 2009, he launched expeditions to various archaeological sites in Peru with the help of the French television station TF1, the city of Toulouse, and private donors. Thierry Jamin published several articles and the book *A la recherche de Paititi*. Paititi is considered the place in Inca mythology from which the last Inca ruler had been taken to heaven. In addition, Paititi is often associated with the fabled golden city of El Dorado, which is a mistake because El Dorado clearly has its origin in Colombia and not in Peru.

Thierry Jamin is considered an adventurous, but absolutely honest, researcher who currently lives in Cusco, Peru. In 2009, he founded the Instituto Inkari (Instituto Inca de Investigacion y Valoracion de Indigena). He is the current president of this organization, which serves as the place of information and connection regarding enigmatic archaeological objects. This is also the reason why grave robbers often show up there with their finds. They hope to find names of possible buyers for their illegally obtained objects.

With little information about the "alien mummies," I was not ready to transfer US $100,000 anywhere. And the image of the "doll in white" did not help. Alien mummies with three fingers and three toes? Somewhat strange, but not impossible. I searched for more information, but did not come across anything really tangible. Then, at the mammoth conference *Contact in the Desert*, held in Palm Springs, California, in 2018, I listened to a talk by Steve Mera and Barry Fitzgerald. I knew Steve Mera as a serious researcher and fellow writer. Steve and Barry had heard of the alien mummies, like I had, and followed their tracks. In their lecture entitled "The Mysterious Bodies of Peru," they described their search for these mummies. Steve and Barry had traveled to Cusco and met Thierry Jamin there. Again Thierry confirmed that the mummies had been discovered in Nazca. And that revived my doubts about the incredible story.

Why does anyone find mummies in Nazca and transport them to Cuzco? Nazca, with the world-famous plain and grandiose lines and removed mountain peaks, lies in the lowlands. Cuzco with its famous Inca fortress, however, lies 3,500 meters high in the Andes. I have already travelled the route once by car, which was a tedious ride through gorges and over mountainsides—a total of 650 kilometers or a thirteen-hour drive. So why were the mummies found in Nazca transferred to Cuzco?

In Cuzco, Steve and Barry were asked to get in a car and not to memorize the make or license number of the vehicle, let alone take a picture. The journey ended in a lousy storage shed, somewhere on the outskirts of Cusco. There they also met "Mario," one of the grave robbers. He reaffirmed that the mummies were only found near one of the Nazca lines because "magnetic irregularities" had been measured in that area. That's true. Years ago, a team of scientists from the Dresden University of Applied Sciences had carried out various analyses in Nazca on behalf of the Erich von Däniken Foundation, and the team detected definite changes in the magnetic field. In the storage shed, Steve and Barry finally saw the white-powdered skeletons of small creatures with long-drawn skulls. Some actually had three fingers and three toes.

But all in all, neither Barry nor Steve were convinced, and the secrecy surrounding the matter made them suspect they would be betrayed. Even more so now, since the real discoverers of the finds did not show themselves and it was unclear where the mummies were right now and who the negotiators would be. In addition, Cuzco's indigenous market offered not only deformed skulls but also small mummified embryos of indeterminate origin. In their presentation, Steve Mera and Barry Fitzgerald warned future researchers not to fall for a phenomenally staged fraud. Meanwhile, they have at least partially revised their opinion. And what about me? Should I continue to care about it, or forget the alleged mummies of aliens?

Life went on. In the back of my mind I had not filed away the story completely. A few months earlier, my colleague Ramon Zürcher and I had led a group to Peru.

Skeletons—Not from This World

The goal wasn't to see the "Inca mummies," but to research other unresolved mysteries in the Andes. The evening before we left for Europe, we were invited to dinner by the commanders of the Peruvian Air Force, Air Force General Luis Spicer Wittembury and Commander Julio Chamorro Flores, whom I knew from previous visits. During dinner, the senior officers reported quite objectively on their experiences with UFOs at a young age. This would never happen in Europe! There I couldn't imagine a German or Swiss air force officer describing, without trepidation, his or her own UFO experiences from earlier years as an active pilot.

At some point, one of the officers asked, "Herr von Däniken, have you ever researched the culture of Chincha?"

"Chincha?" I inquired helplessly. "Who are they?"

"The Chincha are an extinct tribe. They lived about 200 kilometers south of here. Pretty close to Paracas. The Chincha were the people with the three fingers."

Baffled, Ramon and I looked at each other. Our brain cells worked in parallel. The high military we were talking to knew nothing about my search for mummies with three fingers, and now there should be an extinct human tribe with three fingers?

"In what area did these Chincha live?" I inquired, "what do you know about them?" The "Commandante" of the Air Force knew that the tribal area of the Chincha had been the land around Paracas. Today's city Chincha Alta is the sixteenth largest city of Peru and is located only 210 kilometers south of Lima. Did I want to go there?

Ramon and I were scheduled to return to Switzerland the next day. Our flights were booked, but spontaneously we rescheduled. The hunting fever had caught us, and we followed the trail of the three-fingered people. Mr. Julio Chamorro Flores, the commander of the Air Force, made a few phone calls and said that Mr. Giorgio Piacenz, a good friend of his, would accompany us to Chincha Alta in the morning.

After a three-hour drive, Giorgio Piacenz stopped in front of a posh hotel in Chincha Alta. Drinks and appetizers were served in a cooled room, and finally some of the ladies and gentlemen of the provincial government, all elegantly dressed, entered the small hall and greeted us with handshakes and nice words. What was the point of this? Then the lady running the tourist office read a document, which was then solemnly handed over to me. It was titled Resolucion Subperfectural N ° 002-2018-Mininter-Pre-ICA-Sub-Chi-B and dated February 14, 2018. A tribute to me. The province was pleased with my visit and my interest in the culture of the Chincha.

The local archaeologist led us to the outskirts of the city to the ruins of the Chincha Baja. I learned that the culture of Chincha emerged around 800 BC. At first, it settled in what today is Paracas, and then later spread to the coast and down to Nazca. In fact, the Chincha only had three fingers and three toes. This would also be verified through textiles and petroglyphs (rock drawings). Amazingly, so the archaeologist said, was the fact that the Inca rulers had never subjugated the culture of the Chincha. All other peoples who lived in the territory of the

Skeletons—Not from This World

"AÑO DEL DIALOGO Y RECONCILIACION NACIONAL"

RESOLUCION SUBPERFECTURAL
N° 002-2018-MININTER-PRE-ICA-SUB-CHI – B

Chincha Baja, 14 de Febrero del 2018

VISTO:
Que, con fecha 14 de Febrero del año 2018 el escritor Antón Paul Von Daniken, visita el distrito de Chincha Baja –Parador Turístico la Centinela.

CONSIDERANDO:
Que, es política estimular y reconocer a las personas y/o instituciones que destacan en los diferentes campos del quehacer social especialmente cuando son dignos contribuyendo en bien de la comunidad.

Que el distrito de Chincha Baja ha recibido la visita del Escritor Erich Antón Paul Von Daniken, la subprefectura considera pertinente extender un especial saludo y reconocimiento a tan distinguida personalidad que nos honra con su presencia, por su trayectoria como escritor internacional y por el trabajo que viene realizando y difundiendo.

SE RESUELVE:
Artículo Único: Expresar Homenaje de saludo y Reconocimiento al Escritor Erich Antón Paul Von Daniken. Por su visita al distrito de Chincha Baja, Provincia de Chincha, Departamento de Ica. Que el éxito acompañe siempre su noble labor.

REGÍSTRESE, COMUNÍQUESE Y CUMPLASE

Official welcome document of the city of Chincha, Peru

later Inca were forced into serfdom, often in a very bloody way, but not the Chincha. The highest adviser of the highest Inca ruler, I learned, had always been a Chincha. In fact, the Inca palaces were located right next to the Huaca La Centinela, the royal seat of the Chincha.

All ways lead to Rome, a 2,000-year old Roman proverb tells us. All the evidence leads to Nazca I could say today; or at least the evidence in search of the extraterrestrial mummies. The Chincha lived in the area of today's Paracas or Nazca. There are also the famous Nazca lines, which are still not explained correctly in so-called scientific TV documentaries.[1] In Nazca, a tomb with small-sized, possibly extraterrestrial beings is said to have been discovered only because grave robbers used modern equipment to follow magnetic changes. Right; such deviations of the magnetic field had already been published before the grave robbers in two scientific papers: *The Nasca and Palpa geoglyphs: geophysical and geochemical Data*[2] and *Geoscientific View at the Nazca Lines*.[3] The grave robbers did not just search indiscriminately. Presumably, they were familiar with the Spanish version of the treatises on the irregularities in the magnetic field at the Nazca plain. Magnetic changes on the Earth's surface can be easily located with modern magnetometers. In the same geographical area around Paracas lived the Chincha, and they must have been extraordinary people, not only because they had three fingers and three toes, but also because of their intellectual abilities. They were the chief advisors of the Inca rulers.

In addition, countless tombs with deformed skulls have been discovered in the same geographical area. Some of them do not fit into the human line of evolution (see Chapter 1, "Alien Skulls"). There are also ceramics and petroglyphs of people with three fingers. And there is something else: in the Bay of Paracas on a hillside, we find—a 250-meter-high trident called Candelabro, about which I have already speculated.[4] The structure is reminiscent of a gigantic candelabra with three arms. The

Skeletons—Not from This World

Chincha textiles and paintings with three fingered beings

individual columns of this trident are up to 3.80 meters wide. Nobody knows what the structure means, who created it and when it was created.

The logical explanation would be that it is a marker for ships. But even this cannot be true. In front of the coast lies a small island, inhabited by roaring sea lions and other creatures. This island obstructs the view from the Pacific Ocean, and even from north to south the bay of Paracas is only visible for a few kilometers. In addition, the island out at sea would itself be the best marker for the sailors. It can be seen from afar, while the Candelabro appears only after passing the island. In the past, the trident was said to be pointing to the plain of Nazca, a hundred kilometers away. Not correct. The arms of Candelabro do not point to Nazca.

After my new findings about the Chincha, I asked myself whether the Candelabro could have anything to do with the three alien fingers facing the sky looking like the trident of the god Poseidon?

At home, a thick envelope awaited me, which was sent to me by Hans-Peter Jaun, a good friend and librarian. In it, I found a special edition of the journal *Ikaris* published in France.[5] I was baffled. In a detailed fifty-page report, the entire affair of "the extraterrestrial mummies in Peru" was detailed, with many pictures and interviews of the persons involved up until the end of 2017. Thierry Jamin wrote that "Pandora's box was opened in 2016." So I learned through the magazine *Ikaris* how the fantastic story had started. In October 2016, according to Thierry Jamin, an unknown person had come to speak with

Skeletons—Not from This World

The "Candelabro" geoglyph of Paracas

someone at the Instituto Inkari in Cusco. That was not uncommon, because again and again, amateurs would appear and ask for advice about objects.

The stranger, later identified as "Paul R.," showed him (Thierry Jamin) some dried-up organs, but also a hand with three long fingers, a deformed skull, and finally even a whole small body of almost 40 centimeters in length with three fingers on its hands and three toes on its feet. Thierry was shocked and very skeptical. Paul R. had confided to him that the enigmatic find had been made by tomb robbers (in Spanish: *huaqueros*) in Nazca, not far from one of the mysterious lines. Paul R. assured him that he knew the leader of the grave robbers personally, but that he wasn't at liberty to reveal his name. It was only later that Thierry Jamin learned that this Paul R. was a member of a cult in which aliens play the leading role. (To my knowledge, there are only two such communities worldwide: the Rael Movement or Rael Philosophy, headquartered in Canada, and the Church of the Scientology. More can be found on Wikipedia.) This realization did not make the already confusing issue more credible to Thierry Jamin, and he wondered if these unknown grave robbers had actually managed to make the most sensational find in millennia, or if it was a wacky scam. So he kept in contact with his informant Paul R. to learn as much as possible about the place of discovery.

Thierry Jamin finally drove to Nazca and, thanks to the mediation of Paul R., met the as-always-nameless boss of the robbers. He said that the first find had been made in the fall of 2015. They discovered an anomaly in the ground that stood in front of the entrance to a grotto. After overcoming some obstacles, they found a staircase that led to a smaller hall. Two human mummies sat on the ground. (Incas were customarily mummified in sitting position.) From the hall, several paths led to other rooms and closed doors. Two large, sealed sarcophagi stood in a niche. In one of the sarcophagi were around a thousand small, stones carrying engravings, including UFO-like representations. The second sarcophagus contained 20 small, humanoid bodies, all coated with a white powder. For the first time in their life, the astonished grave robbers were speechless. Then they decided to close up the tomb again, for the time being, keep silent about it, and try to figure out whom they should talk to about this.

In 2016 and 2017, the grave robbers were said to have discovered new spaces and objects in the mysterious tombs. Among these were objects that looked completely alien and made no sense to them, including metal spheres and thin glittering foils that always returned to their original shape when scratched, crumpled, or crushed. One of the grave robbers even confessed that he had been able to sell some of these foils to a Japanese man for good money. Throughout the dig, the robbers always had the feeling of being watched by invisible eyes. Two of the grave robbers died shortly afterward of mysterious diseases.

They said they wrapped some of the mummies in cloths and took them to an apartment in Palpa (near Nazca). Finally, the leader of the grave robbers had instructed "Mario" to travel to Cusco and contact the

Instituto Inkari. They couldn't do anything with their finds and needed help.

Slowly Thierry Jamin understood the enormity of the find and got Mario to trust him. Thierry understood that the mummies needed professional protection. They could not be allowed to disintegrate in their new environment and be destroyed by bacteria. In addition, international laboratories had to examine the bodies and get answers to some questions. Which species did they belong to? Did the mummies contain bones stolen from various places? What was the white powder? What would age dating, DNS analysis, and forensic diagnoses determine? Thierry Jamin also knew that the Peruvian state had to be informed about the findings. So, in a detailed letter, he informed the Minister of Culture of Peru, Dr. Nieto Montesinos. He received no answer. Disappointed, he tried the president of the Peruvian Republic with the same result. The high officials did not believe a word of the fantastic story. Only now, after the state authorities had shown no interest did Thierry Jamin contact the *Alien Project*. Funds had to be secured to pay for an international team of scientists. The mummies should be preserved and examined according to all normal protocols.

Only now, because he needed financial support, did Thierry Jamin notify some organizations, such as the National Geographic Society, the well-known Mexican journalist Jaime Maussan, the United States TV station *Gaia*, and, through my acquaintance in Luxembourg, me. Thierry Jamin did not want to become a useful pawn in a scam. He wanted to search for the true origin of these mummies, and if he was able to expose crooks who wanted to fool us all, he would gladly do that, too.

Thanks to the help of broadcasting company *Gaia*, an unbelievable story was about to be told. *Gaia* probably paid tens of thousands of dollars or more for the story. I know the owner of *Gaia*, Mr. Jirka Rysavy, quite well, but I never asked him about his expenses. *Gaia* is headquartered in Boulder, Colorado, USA, and is not a public television service. *Gaia* is Internet TV and therefore accessible to everyone. In 2017, I hosted a multi-part series for *Gaia* on alien visits millennia ago. For weeks, my secretary Ramon Zürcher and I went in and out of the *Gaia* studio in Boulder. We know the "ambience" and some of the people in the executive suite. *Gaia* seeks truths, even if they are uncomfortable. Now *Gaia* was working for or against these Peruvian mummies. Was everything a huge scam or a scary fact?

Gaia found a number of scientists from a variety of specialties and countries who volunteered to examine samples of the mummies and do it completely unbiased, following certain criteria. The silent TV observer and host was Mexican TV journalist Jaime Maussan, winner of the prize "El Sol de México" for investigative journalism. I know Jaime personally. He asserted several times, "I want to find out the truth about these damned mummies. And if a fraudulent gang is behind it, we'll expose it." By mid-2018, *Gaia* had broadcast nine episodes, always including the latest research results.

1. "The Discoveries"

2. "The Excavations"

3. "The Update"

4. "The Evaluations Begin"

5. "The Investigations"

6. "The Little Bodies"

7. "Press Conference with Jaime Maussan"

8. "The Investigations Continue"

9. "The Facts"

All *Gaia* broadcasts were made with the assistance of Ms. Melissa Till, director of original content, and reviewed by Jay Weidner, senior director of content. Each sentence had to agree with the images shot by the camera. There were no anonymous statements. The pictures I show are based on the thorough research and generosity of *Gaia*.

To distinguish the different mummies, they were given human names. There is a "Maria," a "Victoria," a "Josefina" and an "Alberto." The first question for science was in regards to the white powder that covered all the mummies. The stuff turned out to be kieselguhr. What is that? Kieselguhr is a mixture of minerals from dead, microscopically small plants. It contains silicon, magnesium, and calcium and was produced by the pulverized fossil shells of extinct diatoms (rock algae).

These diatoms can be found in both fresh and salt water. Today diatomaceous earth is used as a vermin powder. It has a biophysical effect. Here, knife-sharp edges tear open the chitin shell of crawling insects and arachnids. All pests are dehydrated by diatomaceous earth. It is used against mites, fleas, lice, ticks, cockroaches, ants, silverfish, and snails. It also spreads to the sleeping places of pets. The insects can't build up any resistance to it so, unlike chemical products, it can be used again and again. Therefore, it is also scattered without hesitation in stables or beehives.

Whoever covered mummies with white powder knew its effect. Diatomaceous earth kills all pests and protects the body from decay.

At the press conference on July 17, 2017, Jaime Maussan confirmed, "We have not altered the mummies in any way. If they were fakes, the fraudsters would have studied anthropology at America's Harvard University."

Then Dr. Raymundo Salas who holds a PhD from the University of San Marcos, Lima, and is a radiologist and computer tomography specialist, commented,

> *We examined the mummies thoroughly. The most important thing is their hands and feet. We detected no manipulations. The carpus and metacarpus bones correlate in the hands and in the feet. There are no distortions. The hands have five phalange bones (= joints) and they are very long. The skull is extended backwards. All bones fit together perfectly. There are no manipulations. It is not a normal, human being. Definitely not. This is not an assessment. The diagnoses with CT scan are objective. It's nothing subjective—nothing interpretable. And the mummy is around 1700 years old.*[6]

These findings were confirmed by forensic expert Dr. José de Jesús Zalce Benítez. His team had examined the mummies for errors that counterfeiters might have made. "The tendons and ligaments are perfect. No bones have been glued together. Nothing about it is composed of foreign parts. The anatomy is right. The fingers are about five individual joints. These joints cannot be faked and then glued together. The anatomy on all three fingers is harmonious. Also biomechanically perfect, otherwise the creature would not have been able to make a fist. In addition, the 'foramen magnum' (the occipital hole, exit point of the spinal cord) had shifted. This makes a forgery impossible, because the skeleton consists of one piece, as we can prove on the mummy 'Albert.' All bone parts are normal and organic—but each body has grown in its own individual way. 'Albert' in his natural state of growth. The chin is too small, the ribs are arranged horizontally, the spaces between all bones are perfect. But because the bodies are not all exactly alike, we speak of individuals . . . So the mummy 'Victoria' is similar to that of 'Maria.' But they are not alike."

Dr. Konstantin Korotkov from the University of Saint Petersburg in Russia is also president of the International Union of Medical and Applied Bioelectrography. He personally participated in taking the various samples. These samples were examined by the Federal Research Organization Institute of Obstetrics, Gynecology, and Reproductive Science named after Dr. Otto (Mendeleevskaya line 3, St. Petersburg). Dr. Korotkov explained,

> *The first step was to prove that bones and skin are real, not made of plastic or anything else. Then we needed to determine the age. So we used the C-14 method. Then we analyzed the body structure using computed tomography and[by] X-ray examinations. So it was proved that they are real bodies. But why the three fingers? The three toes? I could take a sample*

Skeletons—Not from This World

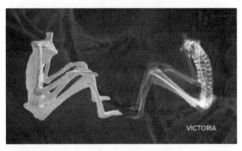

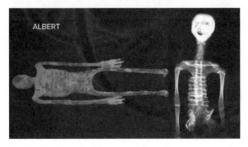

Results from the computer tomographical scans

of one of Maria's fingers and one of her toes. We examined the samples separately again. The most important thing: we were able to prove that the samples are of the same material. This also applies to the same chemical composition. It is the same DNA. Therefore, there is no longer any doubt that the three fingers and the three toes belong to the same body. It had already been proved with the first DNA analysis that it could not be a chimpanzee, but that the DNA was human-like DNA because today we know three types of human-like creatures: Neanderthal, Cro-Magnon, and Denisova hominins. The latter was determined as another type of human because of differences in its DNA. In addition, we know from the CT scan that all internal organs were intact . . . From my point of view, it cannot be a malformation of a normal person. It's an absolutely novel skeletal structure and it belongs to a strange species—another human type.[7]

This was also confirmed by Peruvian radiologist Dr. Raymundo Salas: "We can clearly prove scientifically that these bodies are real bodies. They are not fakes."

Scientists from different institutes who had never met before and performed tests using different methods, confirmed the authenticity of the mummies. For example, the BioTecMol (Moleculares SA) of the Cuidad de Mexico performed DNA analyses. The report from June 24, 2017, signed by Dr. Rogelio Alonso, notes that the DNA is definitely different from human DNA. The DNA had also undergone a purification process at the University of Harrisburg, Pennsylvania, USA. The report, signed by biotechnologist Dr. Doug Taylor, once again notes that the DNA is real and wasn't the same as human DNA. In addition, "Maria" is not a mummified, but a dried-up corpse—all internal organs are present. Dr. William Brown, molecular geneticist at New York University, adds: "I would say that the beings from which the samples were taken have no terrestrial origin known to us. They do not seem to agree with any species of this earth."

Three fingered humanoid mummy—with strange DNA

Dr. Oleg Sergeevich Gotov, senior researcher of the University of St. Petersburg, Department of Molecular Genetics, added: "We were able to extract the ancient DNA... The fact that this ancient DNA has many variants of the common genome is quite startling. In other words, it means that we basically have unique and new data without having added anything."

Dr. José de Jesús Zalce Benitez of the National School of Forensic Medicine of Mexico City stated that it was not a human skeleton. The same was confirmed by Dr. Michael Asseev, head of Genetic Analysis of the Medical Institute, MIBS, of Mexico City.

At the hospital of the University of Colorado, USA, the radiologist M. K. Jesse examined one of the mummies. "The upper and lower jaws show rows of teeth. Thoracic vertebra, thorax and femurs are symmetrical. The skull is slightly elongated, the creature has three fingers and three toes. There are no separate bones. Everything grew together."

Even in far-off Sri Lanka, a gene-tech laboratory, actually a special laboratory for genetic analysis for criminal cases, was involved in the research. Its representatives said that "Maria's" genes could lie between those of a gorilla and those of a chimpanzee. "That would make her a member of an unknown primate group."

Biologist Dr. José de la Cruz Rios, a microscopist at the University of Campeche, Mexico, put it finally as follows: "Since the mummy bears no resemblance to any of the species that we are related to, it is a species very different from us. Why are these beings on our planet if there are no paleontological finds of them? So our conclusion is that the origin of these creatures is not on this planet."

Striking about all mummies was the lack of hair. In addition, a skeleton under the cervical vertebra carried an implant in the form of a "metal strip" under the skin. Whenever the thing was implanted millennia ago, it had to be done at a young age of the life form in question, because the skin had grown seamlessly over it. In order

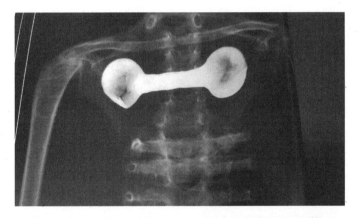

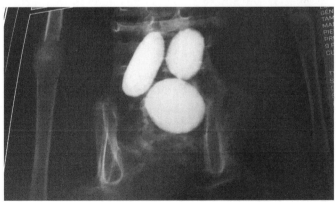

Some of the X-rayed bodies have some kind of eggs in their bodies.

to avoid permanent damage, the metal strip has not been removed to this day. The same mummy had three white eggs in her abdomen, which was even more perplexing. They were clearly visible, and mammals do not lay eggs, birds do—and reptiles.

Let's take a short break. Small creatures with big heads should have come from outer space? Is this crazy? I remembered a report by the (former) Soviet geologist Dr. W. Saizev. He published an article in *Sputnik* magazine in 1968 where he wrote about little beings that had come "from the firmament."[8] The story was later confirmed to me by the Soviet writer Alexander Kasanzev in Moscow. I reported about it in *Zurück zu den Sternen* (*Back to the Stars*).[9] Then my colleague Peter Krassa (1938–2005) examined it, and again later, Hartwig Hausdorf and Jörg Dendl dealt with the matter. [10, 11, 12, 13] Each researcher added a new mosaic to the controversial story or corrected a previous fallacy. Following are the details.

In 1937, a Chinese archaeologist in a mountain cave near Baian-Kara-Ula, located in the Chinese-Tibetan border area, is said to have found 716 tombs with small skeletons.

In the same cave were also stone plates with grooves, only much coarser than those on vinyl records. These grooves contained characters that could later be translated. One told of little beings that came from outer space who called themselves Dropa. Their sky-bound vehicles could not be repaired, so they tried to survive on Earth, but the natives chased them with their horses and killed them.

Did some of these Dropa escape? Did they sail to the Pacific Bay of Pisco-Paracas and become the forefathers of the Chincha? Are the little mummies that were discovered in Nazca their descendants? Were there so many of them because their parents are descended from a reptile species and therefore laid eggs?

I do not know the answers, but I do not want to leave this cross-connection unmentioned. Maybe it can help later researchers.

So far, I have quoted eleven scientists, each a specialist in his field. Others have agreed with them as well, but do not want their names printed. Why? It has to do with chairs at state universities. They fear the mockery by their colleagues, the loss of credibility, or even the fact that the administration of the university knew nothing about their investigations. ("Please! What nonsense are you dealing with? This hurts our reputation . . .") I respect the anonymity of researchers who do not want to expose themselves to the ridicule of their own colleagues. I am well acquainted with the way humans react.

But is there only consent? Where are the contradictions? We know today that the mummies cannot be "normal" humans, but the proof that they actually came from outer space is missing. Although it is quite correct when Dr. Konstantin Korotkov stated that "the skeletal structure belongs to an alien species—another humankind." And biologist José de la Cruz Rios's statement is equally correct, when he points to the lack of paleontological finds of these beings and therefore states that "the origin of these creatures is not on this planet." But the so-called "smoking gun," the solid proof of the alien evolution of this species, is missing. But it could be provided quickly. Come again?

The mummies show completely different ages, varying from around 1700 years to 3600 years. So is it all a hoax?

Dating of the bodies was performed using the so-called C-14 method. This C-14 is a carbon isotope and, according to popular belief, the same amount of it always exists in our atmosphere. The C-14 isotope is slightly radioactive. It comes from the radiation of the universe and is completely harmless. Now, every life form—whether plant, animal or human—absorbs this C-14 every day. Once a living thing dies, it stops absorbing C-14. That's when the time of decay begins—the *half-life*. The very weak radioactive radiation dissolves. For the carbon isotope C-14, this half-life is 5,568 years. After 5,568 years, only half of its quantity or of the original radiation activity is present, and after 11,200 years, only a fourth of it. The method was developed by Frank Libby (1908–1980), then professor of chemistry at Princeton University, New Jersey, USA. In 1960, he received the Nobel Prize for it. But there is a catch with this method. It is assumed that the *same amount* of C-14 isotopes is always present in our atmosphere, but that is not always true. Short fluctuations due to space radiation can corrupt the entire result. Because of such anomalies, a life form would only have to take up a little bit more C-14 than expected, and the dating would provide completely different data. Also, if the Peruvian mummies are *extraterrestrial* bodies, their metabolism could absorb different amounts of C-14 than we earthlings absorb. Our measurements are based on C-14 levels in the *terrestrial* environment. We have no idea which atmospheres those beings came into contact with. Perhaps they also worked with slightly radioactive objects or flew through the radioactive Van Allen Belt in our solar system. There are twenty possible explanations of why the C-14 method gave different results for the Peruvian mummies.

In one of the sarcophagi in the dark tomb, there are many small, engraved stones, claimed one of the grave robbers. The collection of the late Dr. Janvier Cabrera (1924–2001) in Ica, Peru, shows exactly the same: masses of engraved smaller and larger stones. The town of Ica, which houses the Cabrera collection, is located right in the middle between Paracas and Nazca. Dr. Cabrera, whom I knew well, could distinguish between ancient and modern engravings. He had personally taken me to a forger who demonstrated how today's copies are made. It had also been Cabrera who had commissioned several scientific assessments of the stone carvings. I asked pressing questions as to where the masses of engraved stones came from, and he always responded with a secret locality—"the depot." *And we knew all this already 40 years ago.* (I wrote about it repeatedly.[14, 15]) Hundreds of engraved stones contained in a sarcophagus in the same region cannot shake my convictions.

Foils were also discovered in the tombs at Nazca that, when crumpled up, in other words after their shape had been altered, always returned back to their original shape as if straightened out by an invisible hand. Was this a miracle? Foils with the same characteristics have been secured as early as 1947 after the ominous UFO crash in Roswell, New Mexico. The newspapers of that time talked about it, and today, the pictures can be admired at the UFO Museum of Roswell. In addition, stainless metals and thin foils were already described in the *Hitat*.[16] This creates a connection with the treasures in the Great Pyramid thousands of years ago. How were these special foils, also called "intelligent metal," at a subterranean site? And why were they always in connection with extraterrestrial mummies? It doesn't bother me; I have heard it many times before. But didn't the media a decade ago already report about a mummy discovered in the Chilean Atacama Desert? The body, which is just 15 centimeters long, has since been identified as human. This was confirmed by studies carried out at the University of California at San Francisco as well as at Stanford University in Palo Alto. It was a human deformity—nothing more. Could not the Peruvian mummies also be deformities? This explanation does not work: the "Peruvians" are not human, but originally reptilian. Remember the eggs in the body of one of the mummies. So, are we talking about some stupid creatures that the earth was teeming with? No, because the stupid creatures must have actually been intelligent, or otherwise they would not have rubbed their dead bodies with diatomaceous earth and erected tombs with doors and sarcophagi, let alone all their technological know-how.

Something else bothers me: the indifference of the media and the total lack of interest of the authorities. It may take years before this exciting find is finally discussed in major magazines. Then they will come running—the journalists and the offended scientists—and it will be too late. Today one can already read on the Internet about a trade with the mummies of Nazca and with the artifacts found there.[17] There are rumors of several subterranean spaces, objects being hijacked for 20,000 and more US dollars, and a black market price of over a million dollars for a mummy. And we read about "powerful and greedy forces, who are opposed to uncovering true history and human evolution."

Is this all just a new conspiracy theory? The facts are on the table. The mummies exist; after all, they were x-rayed and examined with the help of computed tomography. DNA analyses of several institutes are available. Denying the existence of the alien bodies is no longer possible. But by the summer of 2018, expert investigator Thierry Jamin from Cusco no longer knew who had the mummies. There are already films appearing on the internet that discredit the whole story about these mummies and ridicule them. It's exactly as I said in the beginning of this chapter: "Smart alecks will call the whole thing fake, others a targeted fraud." It's no use. *Gaia* has done their job as professionally as possible.

EPILOGUE

And Yet It Moves!

HOW LONG DOES IT TAKE for any previously controversial opinion to become common knowledge? Will it last until the so-called zeitgeist changes? *Science Online* published a study on it.[1] The question was: under what conditions does an opinion prevail that does not conform to the current conventions?

The answer, in terms of quantity, is that just 25 percent of a group will be enough. Then the opinion will spread rapidly. After a few years, 70–100 percent of people will be convinced of the new view. On the question, "Did the Earth receive visitors from space millennia ago?" the overturning of the opinion has long been in progress. Counting my fellow writers and myself, the worldwide edition of books dealing with this very question is over 200 million copies strong. Add to that thousands of articles in newspapers and magazines and hundreds of TV shows. Above all, the *History Channel* in the US with its (so far) 180-episode strong series *Ancient Aliens*. And *Gaia* TV. It's no use if a religious group tries to block the new findings. Those ancient scriptures with the knowledge of our forefathers still exist. Thousands of years ago, they were hidden with the intention of *opening the eyes of future generations.* What was that again about the construction of the Great Pyramid before the Flood? Why was it built? "... *All the secret sciences were recorded . . . and all other the sciences . . .*" And why did Enoch hand over to his son Methuselah the books he had written and requested "*to save them for future generations living after the Flood . . .*"? For that reason. And they will be discovered again. And what if they are not?

Meanwhile, we have photographed the lunar surface and the surfaces of Mars and the great asteroid Ceres. The images everywhere show clear traces of intelligence: rectangles and circles in the ground, straight lines like lanes on the moon and obelisk-like towers on the Martian moon Phobos. Since we humans have visited no other celestial body except the moon, the traces can only come from an alien intelligence. Despite diligent censorship on the part of the diehards, the findings can't be stopped. Pandora's box can no longer be closed, and the information escaping from it will pass through the thickest monastery walls. The message of aliens is also in our genes and was passed down from generation to generation with the ineradicable pleasure of sexuality.

NOTES

CHAPTER 1

1 Pollak, Kurt: *Wissen und Weisheit der alten Ärzte*. Düsseldorf, 1968.
2 Khudaverdyan, A.: "Cranial deformation and torticollis of an early feudal burial from Byurakn, Armenia." In: *Acta Biologica*, Nr. 56, 2, 2012.
3 Wagner, Gernot: *Die künstlich deformierten Schädel von Österreich in der Frühgeschichte*. Diplomarbeit. Universität Wien.
4 Hausdorf, Hartwig: *Götterbotschaft in den Genen*. München, 2012.
5 Dingwall, E.J.: *Artificial Cranial Deformation*. London, 1931.
6 Virchow, Rudolf: "Über Schädelform und Schädeldeformation." In: *Correspondenz-Blatt der Deutschen Gesellschaft für Anthropologie, Ethnologie und Urgeschichte*. Berlin, 1892.
7 Schröter, Peter: "Zur beabsichtigten künstlichen Kopfumformung im völkerwanderungszeitlichen Mitteleuropa." In: *Die Bajuwaren*. Arbeitsgruppe Bajuwarenausstellung, 1988.
8 Abarzua/Posselt: "Gräber aus uralter Zeit: Tote von andern Sternen." In: *Bild*, 29. April 1975.

9   Lizárraga, Reginaldo de: *Descriptión y problación de las Indias.* Lima, Peru, 1908.

10 Cieza de León, Pedro: *Crónica del Perú.* Sevilla, 1552.

11 *Die Heilige Schrift des Alten und des Neuen Testaments.* Stuttgart, 1972.

12 Berdyczewski, M.J. (Bin Gorion): *Die Sagen der Juden.* Band 1: *Von der Urzeit.* Frankfurt/M. 1913.

13 Kautzsch, Emil: *Die Apokryphen und Pseudepigraphen des Alten Testaments.* 2. Band: *Buch Henoch.* Tübingen, 1900.

14 *Die Argonauten des Apollonius.* Zürich, 1779.

15 Mooney, G. W.: *The Argonautica of Apollonius Rhodius.* Dublin, 1912.

16 Delage, Emile and Vian, Francis: *Apollonius de Rhodes.* Tome III, Chant IV. Paris, 1982.

17 Burckhardt, G.: *Gilgamesch, eine Erzählung aus der alten Welt.* Wiesbaden, 1958.

18 Freuchen, P.: *The Book of the Eskimos.* Greenwich, 1961.

19 Bezold, Carl: *Kebra Negast. Die Herrlichkeit der Könige.* München, 1905.

20 Däniken, Erich von: *Götterdämmerung.* Rottenburg, 2010.

21 Weidenreich, F.: *Apes, Giants and Man.* Chicago, 1947.

22 Spörri, Gregor: *Lost God. Das Jüngste Gericht.* Magden, Schweiz, 2017. (In English: *Lost God. Day of Judgment.* Switzerland, 2018.)

23 Bürgin, Luc: "Die Monsterkralle von Bir Hooker." In: *Mysteries,* Nr. 2/2012.

24 Newman, Hugh: "The Giants of Ancient Egypt." In: *Nexus,* Band 25, December 2017.

25 Emery, Brian, W.: *Archaic Egypt. Culture and Civilisation in Egypt Five Thousand Years Ago.* New York 1974.

26 "Fred—der älteste Amerikaner." In: *Bild der Wissenschaft*, August 1973.
27 Däniken, Erich von: *Falsch informiert!*. Rottenburg, 2007.
28 Habeck, Reinhard: *Ungelöste Rätsel*. Wien, 2015.
29 Däniken, Erich von: *Reise nach Kiribati*. München 1983.
30 Däniken, Erich: *Die Augen der Sphinx*. München, 1992.
31 Hausdorf, Hartwig: *Götterbotschaft in den Genen*. München, 2012.
32 Childress, David Hatcher and Foerster, Brien: *The Enigma of Cranial Deformation*. Kempton, Illinois, USA, 2012.
33 Kulke, Ulli: "Der neue Nachbar aus der Höhle." In: *Welt am Sonntag*, Nr. 13, 28. March 2010.
34 "Alternative Geschichte des Lebens." In: *Die Welt*, 9. April 2018.
35 Marzulli, L.A.: *The Expert Analysis. DNA-Results*, 2018.
36 Foerster, Brien: *Elongated Skulls of Peru and Bolivia: The Path of Viracocha*. USA, 2015.
37 Foerster, Brien: *Beyond the Black Sea*. Middletown, Delaware, USA, 2018.
38 Däniken, Erich von: *Botschaften aus dem Jahr 2118*. Rottenburg 2016.

CHAPTER 2

1 Herodot: *Historien*, griechisch—deutsch. II. Buch. München, 1963.
2 Bifiger, Beat und Stanglmeier, Lothar: *Der Kopf des Osiris*. Unveröffentlichte Recherche für die Erich-von-Däniken-Stiftung.
3 Naville, Edouard: *The Cemeteries of Abydos*. London, 1914.
4 Däniken, Erich von: *Der Mittelmeerraum und seine mysteriöse Vorzeit*. Rottenburg, 2012.

5   Jeremis, Alfred: *Die außerbiblische Erlösererwartung.* Leipzig, 1927.
6   Sethe, Kurt: *Übersetzung und Kommentar zu den altägyptischen Pyramidentexten.* Band II. Darmstadt, 1922.
7   Königsliste von Abydos (Sethos I.). *https://de.wikipedia.org.*
8   Spielberg, Wilhelm: *Die Glaubwürdigkeit von Herodots Bericht über Ägypten im Lichte der ägyptischen Denkmäler.* Heidelberg, 1926.
9   James, Preston E. and Martin, Geoffry J.: *All Possible Worlds.* New York, 1972.
10 Karst, Josef: "Eusebius Werke," V. Band. In Die Chronik aus dem Aramäischen übersetzt. Leipzig, 1911.
11 Schwab, Gustav: *Sagen des klassischen Altertums.* Heidelberg, 1972.
12 Helck, Wolfgang: *Untersuchungen zu Manetho und den ägyptischen Königslisten.* Berlin, 1956.
13 Krannich, H. Paul: *Henochs Uhr.* Norderstedt, 2009.
14 Diodor von Sizilien: *Geschichts-Bibliothek.* Stuttgart, 1866.
15 Stuart, David and George: *Palenque, Eternal City of the Maya.* London, 2008.
16 Brasseur de Bourbourg, Charles-Etienne: *Histoire des nations civilisées du Mexique et de l'Àmerique Centrale.* Tome I–VI. Paris, 1857–1860.
17 Apelt, Otto: *Platon—Sämtliche Dialoge. Kritias und Timaios.* 1922.
18 Fiebag, Johannes und Peter: *Die geheime Botschaft von Fatima.* Tübingen, 1986.
19 Parkinson, Richard: *The Rosetta Stone.* London, 2005.
20 Meyer, Eduard: *Ägyptische Chronologie.* Berlin, 1904.
21 Beckerath, Jürgen von: *Chronologie des pharaonischen Ägypten.* Mainz, 1997.
22 Beckerath, Jürgen von: *Handbuch der ägyptischen Königs-namen.* Mainz, 1999.

23 Diodor von Sizilien: *Geschichts-Bibliothek*. 1. Buch. Stuttgart, 1866.
24 Gaius Plinius Secundus: *Die Naturgeschichte*. 36. Buch. Leipzig, 1882.
25 Telex Reuters und SDA vom 16. April 1993.
26 "Secret Chamber May Solve Pyramid Riddle." In: *The Times*, 17. April 1993.
27 Marchant, Jo: "Cosmic-Ray Particles Reveal Secret Chamber in Egypt's Great Pyramid." In: *Nature*, 6. November 2017.
28 Däniken, Erich von: "Veil Dance Around the Pyramids." In: *Sagenhafte Zeiten*, Nr. 1/2018. 20. Jg.
29 "Archäologie | Chephren-Pyramide, Fluch des Pharaos." In: *Der Spiegel*, Nr. 33, 1969.
30 Bauval, Robert und Gilbert, Adrian: *The Orion Mystery*. London, 1994.
31 Cayce, Edgar Evans: *Edgar Cayce on Atlantis*. 1968.
32 Dormion, Gilles und Goidin, Jean-Patrice: *Kheops. Nouvelle Enquete*. 1986.
33 Dormion, Gilles und Goidin, Jean-Patrice: *Nouveaux Mystères de la Grande Pyramide*. 1987.
34 Yoshimura, Sakuji: *Non-destructive Pyramid Investigation by Electromagnetic Wave Method*. Waseda University. Tokyo, 1987.
35 Zarei, Alireza: *Die verletzte Pyramide*. Groß-Gerau, 2011.
36 "Sphinx Riddle Put to Rest?" In: *Science*, Vol. 255, Nr. 5046, 14. February 1992.
37 West, J.A.: *Serpent in the Sky*. 1993.
38 Telex Reuters und SDA vom 16. April 1993.
39 "Secret Chamber May Solve Pyramid Riddle." In: *The Times*, 17. April 1993.

40 Marchant, Jo: "Cosmic-Ray Particles Reveal Secret Chamber in Egypt's Great Pyramid." In: *Nature*, 2. November 2017.
41 Krüger, Paul-Anton: "Leere in der Pyramide." In: *Süddeutsche Zeitung*, 3. November 2017.
42 Speicher, Christian: "Ein mysteriöser Hohlraum in der Che-ops-Pyramide." In: *Neueste Zürcher Zeitung*, 3. November 2017.
43 Imhasly, Patrick: "Des Pharaos neueste Rätsel." In: *Neueste Zürcher Zeitung am Sonntag*, 5. November 2017.
44 Creighton, Scott: "The great pyramid Fraud Revisted." In: *Nexus*, Vol. 21, Nr. 4/2014.
45 Kees, Hermann: *Kulturgeschichte des Alten Orients. Ägypten*. München, 1965.
46 Eggebrecht, Eva: "Die Geschichte des Pharaonenreiches." In: Eggebrecht et al.: *Das alte Ägypten*. München, 1965.
47 *Das Pyramidenkapitel in Al-Makrizis* "Hitat." Übersetzt von Dr. Erich Graefe. Leipzig, 1882.
48 Däniken, Erich: *Falsch informiert!*. Rottenburg 2007
49 Tompkins, Peter: *Cheops*. Bern, 1975.
50 Kautzsch, Emil: "Das Buch Henoch." In: *Die Apokryphen und Pseudepigraphen des Alten Testaments*. Band II. Tübingen, 1900.
41 *Das Pyramidenkapitel in Al-Makrizis* "Hitat." Übersetzt von Dr. Erich Graefe. Leipzig, 1882.
52 Al-Mas'udi: *Bis zu den Grenzen der Erde*. Tübingen/Basel, 1978.
53 Kautzsch, Emil: "Das Buch Henoch." In: *Die Apokryphen und Pseudepigraphen des Alten Testaments*. Band II. Tübingen, 1900.
54 Däniken, Erich von: *Die Augen der Sphinx*. München, 1989.
55 Appelt, Otto: *Platon—Politikos*. Hamburg, 1988.

56 Voss, Johann H.: *Hesiods Werke*. Wien, 1817.
57 Velikovsky, Immanuel: *Welten im Zusammenstoß*. Frankfurt, 1978.
58 Däniken, Erich von: *Der Mittelmeerraum und seine mysteriöse Vorzeit*. Rottenburg, 2012.

CHAPTER 3

1 Velikovsky, Immanuel: *Welten im Zusammenstoß*. Frankfurt, 1978.
2 Lovell, Bernard und andere: *Weltraumatlas*. Bern, 1970.
3 Ruppe, Harry: *Die grenzenlose Dimension Raumfahrt*. B and 2. Düsseldorf, 1982.
4 Clube, Victor und Napier, Bill: *The Cosmic Serpent—A Catastrophist View of Earth History*. London, 1982.
5 Horedt, G.P.: "Origin of the Asteroids." In: *Icarus*, Nr. 22/1974/
6 Frauenholz, A.: *Das Sonnensystem in der Vorzeit. Eine wissenschaftliche Abhandlung*. Breslau, 1869.
7 Gerland, Georg: *Der Mythus von der Sintflut*. Bonn 1912
8 Riem, Johannes: *Die Sintflut in Sage und Wissenschaft*. Hamburg, 1925.
9 Anderson, Walter: *Nordasiatische Flutsagen*. Dorpat .1923.
10 Müller, Werner: *Die ältesten amerikanischen Sintfluterzählungen*. Dissertation. Bonn, 1930.
11 Hohenberger, A.: *Die indische Flutsage und das Matsyapurana*. Leipzig, 1930.
12 Paulson, Ivar: *Die Religionen Nordeurasiens und der amerikanischen Arktis*. Stuttgart, 1962.
13 Frazer, James George: *Folklore in the Old Testament*. London, 1919.

14 Velikovsky, Immanuel: *Welten im Zusammenstoß.* Frankfurt, 1978.
15 *Die Heilige Schrift des Alten und des Neuen Testaments.* Württembergische Bibelanstalt. Stuttgart, 1972.
16 Breuer, Reinhard: "Schwarzes Loch im Zentrum der Milchstraße." In: *Bild der Wissenschaft.* November 1977.
17 Kautzsch, Emil: *Die Apokryphen und Pseudepigraphen des Alten Testaments.* Band II: *Das Buch Henoch.* Hildesheim, 1962.
18 Däniken, Erich von: *Botschaften aus dem Jahr 2118.* Rottenburg, 2016.
19 Kautzsch, Emil: *Die Apokryphen und Pseudepigraphen des Alten Testaments.* Band II: *Das Buch Henoch.* Hildesheim, 1962.
20 "Können wir noch Christen sein?" In: *Der Spiegel,* Nr. 13, 1994.
21 Däniken, Erich von: *Habe ich mich geirrt?* München, 1991.
22 Roy, Chandra Protap: *The Mahabharata, Drona Parva.* Kalkutta, 1888.
23 Roth, Rudolf von: "Der Mythos von den fünf Menschengeschlechtern bei Hesiod und die indische Lehre von den vier Weltaltern." In: *Verzeichniss der Doctoren, welche die philosophische Facultät der Königlich Württembergischen Eber-hard-Karls-Universität in Tübingen im Decanatjahre 1858–1859 ernannt hat.* Tübingen, 1860.
24 Schwab, Gustav: *Sagen des klassischen Altertums.* Heidelberg 2011.
25 Apelt, Otto: *Platon. Sämtliche Dialoge. Kritias und Timaios.* 1922 (Neuauflage Hamburg, 1988).
26 Engelhardt, Wolf von: *Phaetons Sturz—ein Naturereignis?* Sitzungsberichte der Heidelberger Akademie der Wissenschaften. Heidelberg, 1979.
27 Pritchard, James, B.: *Ancient Near Eastern Texts Relating to the Old Testament.* Princeton, 1955.

Notes

28 Eberhard, Wolfram: *Lokalkulturen im alten China.* Teil 2. Leiden, 1942.

29 Gaius Plinius Secundus: *Naturgeschichte,* 2. Buch, Abschnitt 91.

30 Blavatsky, Helena P.: *Die Geheimlehre.* Band I: *Kosmogenesis.* Den Haag o.J 1999.

31 Kanjilaal, Kumar Dileep: *Vimanas in Ancient India.* Kalkutta, 1985.

32 Ganguli, Mohan Kisari: *The Mahabharata.* Vol. III. New Delhi, 2000.

33 Förstermann, Ernst: "Blatt sechzig der Dresdner Maya-Handschrift." In: *Das Weltall.* 6. Jg., Heft 16, Berlin, 1906.

34 Rowan-Robinson, Michael: "Maya-Astronomie." In: *New Scientist,* 18. October 1979.

35 Morley, Sylvanus Grisworld: *The Ancient Maya.* Standford 1946. Und Morley, Sylvanus Grisworld: *La Civilisation Maya.* Mexico, 1947.

36 Hawkins, J.: "An Account of the Discovery of a Very Ancient Temple on Mount Ocha in Euboea." In: *Travels in Various Countries.* London, 1820.

37 Ulrichs, Heinrich: "Über den Tempel der Juno auf dem Berg Ocha bei Carystos." In: *Reisen und Forschungen in Griechenland.* Berlin, 1863.

38 Theodossiou, E.: "Study and Orientation of the Mount Oche Dragon House in Euboea." In: *Journal of Astronomical History and Heritage.* Vol. 12, 2009.

39 Kalaougas, Vasilis: *Die Drachenhäuser von Euböa.* 1998.

40 Däniken, Erich von: *Die Steinzeit war ganz anders.* München, 1991.

41 Däniken, Erich von: *Aussaat und Kosmos.* Düsseldorf, 1972.

42 Däniken, Erich von: *Falsch informiert!.* Rottenburg, 2007.

43 Habeck, Reinhard: *Geheimnisvolles Österreich.* Wien, 2006.

44 Austria-Forum.

45 Habeck, Reinhard: *Geheimnisvolles Österreich.* Wien, 2006.

46 Austria-Forum.

47 Habeck, Reinhard: *Überirdische Rätsel.* Wien, 2016.

48 Bürgin, Luc: *Lexikon der verbotenen Geschichte.* Rottenburg, 2018.

49 Bürgin, Luc: *Chinas mysteriöses Höhlenlabyrinth.* Rottenburg, 2013.

50 Bürgin, Luc: *Chinas mysteriöses Höhlenlabyrinth.* Rottenburg, 2013.

51 Bürgin, Luc: "Höhlen-Labyrinth in Huangshan. Noch größer als vermutet." In: *Mysteries,* Nr. 6/2014.

52 Bürgin, Luc: *Chinas mysteriöses Höhlenlabyrinth.* Rottenburg, 2013.

53 *Wikipedia.* Liste der EinschlagCrater der Erde. *https://de.wikipedia.org.*

54 "Dating the Lybian Desert Silica-Glass." In: *Nature,* Nr. 170, 1952.

55 Farkas, Viktor: "Rätselhafte Verglasungen um den Globus." In: *Sagenhafte Zeiten,* Nr. 5, 2000.

56 Bude, Friedrich: Brief an Erich von Däniken vom 26. March 2018.

57 Biglino, Mauro: *Kamen die Götter aus dem Weltall?* Rottenburg 2015. Seite 88 ff.

CHAPTER 4

1 Däniken, Erich von: *Unmögliche Wahrheiten.* Rottenburg 2013. 5.

2   Hartsch, Kerstin und andere: "The Nasca and Palpa Geo-glyphs: Geophysical and Geochemical Data." In: *Naturwissenschaften,* October 2009.

3   Rosas, Silvia: *Geoscientific View at the Nazca Lines.* Universidad Católica del Perú, Lima, 2013.

4   Däniken, Erich von: *Grüße aus der Steinzeit.* Rottenburg, 2010.

5   "Le Dossier du mois: L'Enigme des Momies à trois Doigts Découvert au Pérou." In: *Ikaris,* Nr. 1, February/March 2018.

6   Salas, Raymundo bei der Pressekonferenz von Jaime Maussan vom 17. Juli 2017. In: *Gaia | Unearthing Nazca.*

7   Korotkov, Konstantin in: Gaia | *Unearthing Nazca.* Update Nr. 6. Boulder 2017. Und Korotkov, Konstantin in: Gaia | *Unearthing Nazca.* Update Nr. 5. Boulder, 2017.

8   Saitzew, W. "Wissenschaft oder Phantasie?" In: *Sputnik,* Nr. 1, 1968.

9   Däniken, Erich von: *Zurück zu den Sternen.* Düsseldorf 1969. Seite 170 ff.

10 Krassa, Peter: *Als die gelben Götter kamen.* Wien, 1973.

11 Hausdorf, Hartwig: "Auf Göttersuche in China." In: *Sagenhafte Zeiten,* Nr. 4, 1994.

12 Hausdorf, Hartwig: "Baian-Kara-Ula. Neue Erkenntnisse zum Jahrhunderträtsel." In: *Sagenhafte Zeiten,* Nr. 6, 1995.

13 Dendl, Jörg: "Die Steinscheiben von Baian-Kara-Ula: Der erste Bericht." In: *Sagenhafte Zeiten,* Nr. 1, 1996.

14 Däniken, Erich von: *Beweise.* Düsseldorf 1977. Seite 411.

15 Däniken, Erich von: *Zeichen für die Ewigkeit.* Taschenbuchausgabe. München 1997, Seite 67 ff.

16 *Das Pyramidenkapitel in Al-Makrizis "Hitat."* Übersetzt von Dr. Erich Graefe. Leipzig, 1882.
17 *https://dieunbestechlichen.com.*

EPILOGUE

1 Dengler, Roni: "How Minority Viewpoints Become Majority Ones." *http://www.sciencemag.org.*

IMAGE REFERENCES

Image 19: Animated graphic by Selina Rüegg, Horgen, Switzerland

Image 24—32: Ramon Zürcher, Beatenberg, Switzerland

Image 44: Reinhard Habeck, Vienna, Austria

Image 45: Dr. Heinrich Kusch, Graz, Austria

Image 46–50: Luc Bürgin, Jenny and Floyd Varesi, Basel and Gelterkinden, Switzerland

Image 58/72/74/75: Thierry Jamin, Alien Project, Cuzco, Peru

Image 68/70/71: Steve Mera, Manchester, Great Britain

Image 69/73: Gaia-TV—Press conference June 20, 2017

All other images: from the Erich von Däniken Archives

ABOUT THE AUTHOR

ERICH VON DÄNIKEN is arguably the most widely read and most copied nonfiction author in the world. He published his first (and best-known) book, *Chariots of the Gods,* in 1968. The worldwide bestseller was followed by forty more books, including the recent bestsellers *Eyewitness to the Gods, The Gods Never Left Us, Twilight of the Gods, History Is Wrong, Evidence of the Gods, Remnants of the Gods,* and *Odyssey of the Gods.* His works have been translated into twenty-eight languages and have sold more than sixty-five million copies. Several have also been made into films. Von Däniken's ideas have been the inspiration for a wide range of television series, including the *History Channel*'s hit *Ancient Aliens.* His research organization, the AAS RA (Archaeology, Astronauts and SETI Research Association), comprises laymen and academics from all walks of life (see *legendarytimes.com*). Internationally, there are about ten thousand members. Erich lives in Switzerland but is an ever-present figure on the international lecture circuit, traveling more than one hundred thousand miles a year.